BASIC NEUROANATOMY

BASIC NEUROANATOMY

By

Carlton George Smith

B.A., M.SC., M.D., PH.D.

*Professor of Anatomy, in charge of Neuroanatomy,
University of Toronto*

UNIVERSITY OF TORONTO PRESS

Copyright, Canada, 1961, by
University of Toronto Press
Printed in Canada
Reprinted, 1962
Reprinted with alterations, 1964
Reprinted in 2018
ISBN 978-1-4875-7320-1 (paper)

Preface

STUDENTS OF neuroanatomy, particularly medical students, are interested primarily in neural pathways, their interrelationships, and their blood supply. These pathways can be charted in diagrams, but diagrams cannot show the changes in the relationships of the pathways to each other, or their relationship to nutrient blood vessels. Hence a three-dimensional concept of the brain and cord is a basic requirement.

To help achieve this three-dimensional concept, each part of the brain and the cord is described as having the geometrical form which it most resembles—cylinder, cube, wedge, and so on—and those features of its surface that are difficult to understand and visualize are interpreted in terms of development. The fact that the brain and cord are made up entirely of closely interwoven pathways—nothing more—is kept to the fore to stress the relationship of form and function.

In this text the caudal end of the axial nervous system is described first because it is here that the longest constituents of the ascending pathways enter the central nervous system and because it is here that the longest constituents of the descending pathways leave it. The successively higher segments of the central nervous system are then described in turn, relating the changes in form and size at each level to (1) the entrance or departure of pathways, (2) the change in position of the pathways relative to each other, and (3) the interpolation of relay stations in these pathways.

In order that the structure of the brain and the course of the longer pathways may be described with a minimum of digression, a full description of the cranial nerves and their afferent and efferent pathways is delayed until after the whole of the axial nervous system has been described.

To round out this study of the nervous system an attempt is made to interpret the structures that comprise its autonomic division. The concluding chapter, fittingly, is a treatise on the arterial blood supply. Because of the importance of an understanding of the blood supply of the brain and cord in making diagnoses, the vessels are described in detail and special dissections were prepared by the author as a basis for the illustrations.

In keeping with the primary purpose of this text, a description of the coverings of the brain and the histology of nervous tissue are not included. This is justifiable because the meninges are studied in gross anatomy and the cells of the nervous system are studied in histology. In conclusion, it is hoped that the indexing of the successive parts of each pathway and each tract in serial order rather than alphabetically will serve the purpose of a tabular summary in each case, and facilitate their review.

Acknowledgments are due to my mother, Mrs. Catherine Smith, for her assistance in all phases of this project but particularly for her help with the illustrations and in preparing the index. In the course of the preparation of this text the author was continually reminded of his indebtedness to Dr. E. Horne Craigie of the University of Toronto, for his early guidance in the field of neuroanatomy, and also to Dr. James W. Papez, at Cornell University, to Dr. Elizabeth Crosby at the University of Michigan, and to Dr. Fred A. Mettler of Columbia University for laboratory privileges and opportunities for research.

CARLTON G. SMITH

Contents

PREFACE v

1. Introduction 1
2. The Spinal Cord 13
3. The Medulla Oblongata 30
4. The Pons Segment 40
5. The Midbrain 49
6. The Diencephalon 59
7. The Cranial Nerves 100
8. The Reticular Formation 123
9. The Cerebellum 127
10. The Cerebral Hemisphere I: Development, Form, and Structure 154
11. The Cerebral Hemisphere II: The Sulci, Gyri, and Cortical Areas 171
12. The Cerebral Hemisphere III: Internal Structure 190
13. The Autonomic Nervous System 207
14. The Blood Supply of the Brain and Spinal Cord 222

INDEX 247

BASIC NEUROANATOMY

1. Introduction

THE NERVOUS SYSTEM is the communication system of the body. It serves to convey excitation from the sense organs—the listening posts of the body—to the effectors—the muscles and glands. When several sense organs are excited simultaneously the nervous system has the additional job of processing the data obtained and selecting the particular effectors that will ensure an appropriate response. The processing of the data is done in a central, axial, portion of the nervous system called the central nervous system. The cable-like structures connecting the central nervous system with (a) the sense organs and (b) the effectors, are the sensory and motor nerves respectively. These make up the peripheral nervous system.

The wires in the cable-like nerves are the nerve fibres. Each of these is a filament of cytoplasm, a process of a nerve cell. The cell body and its process, the nerve fibre, together comprise the anatomical unit of the nervous system called a neuron.

I. THE NEURON, THE ANATOMICAL UNIT OF THE NERVOUS SYSTEM

A neuron, or nerve cell, like all cells has a nucleus and a covering of cytoplasm. Its surface is specialized to detect changes in its environment and accordingly it has a very thin cell membrane. The detecting surface of the cell is usually increased by putting forth many branching, feeler-like, processes. These are called **dendrites.** The nerve cell also puts forth one process that is long and slender and is insulated from its environment except at its tip. This is its **axon** and it conducts excitation away from the cell body. Thus the nerve cell can be divided, functionally, into two portions: a receptive portion, the cell body with its cytoplasmic processes that are not insulated, and a conductor portion, the axon, which is insulated from its environment except at its termination (figure 1).

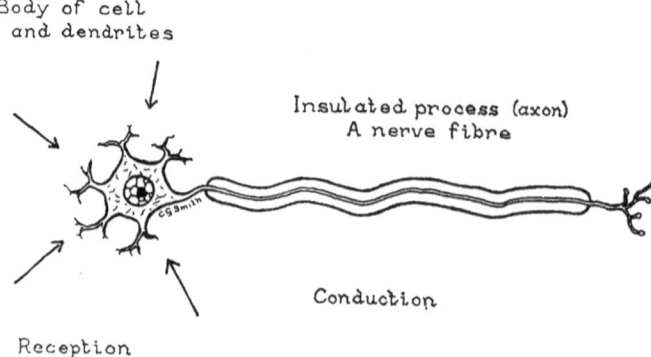

FIGURE 1. The parts of a neuron. The neuron has a part for reception and a part for conduction of excitation.

For convenience in describing pathways in the nervous system we will call a process that is insulated a **nerve fibre**, not an axon. This will be helpful because there are some exceptional nerve cells (sensory nerve cells) that have two long insulated processes, one of which does not qualify as an axon because it conducts excitation toward the cell body, not away from it.

II. THE ARRANGEMENT OF THE NERVE CELLS WITHIN THE NERVOUS SYSTEM. THE FUNCTIONAL UNIT

The neurons are arranged in chains (figure 2). The end of the nerve fibre (axon) of one neuron touches the cell body of another neuron, and so on. The contact is called a **synapse**. Such a chain of neurons serves as a conductor of excitation. The excitation is called a **nerve impulse**. When a nerve impulse arrives at the end of a nerve fibre, and the conditions are right, there will be a transfer of excitation to the cell body it touches.

In higher animals the sense organs and effectors are always connected by a chain of two or more neurons. Hence the functional unit of the nervous system is a chain of neurons. In this functional unit the synapse has two important functions: (1) It acts as a valve permitting excitation to travel in one direction only. (2) It serves as an entrance to the functional unit for impulses from other nerve fibres. These other impulses may facilitate or block the flow of excitation in the functional unit. The synapse is therefore the anatomical device which gives the nervous system its capacity to select effectors, that is, to initiate a response appropriate to the entire environmental situation.

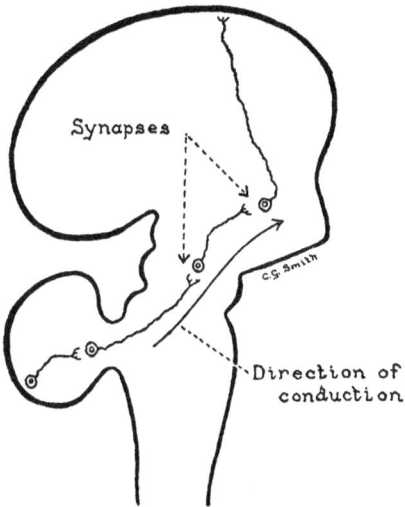

FIGURE 2. An example of a functional unit of the nervous system—a chain of nerve cells.

It was stated above that the neuron has a cell body and one nerve fibre. This is true of all neurons except the neuron that gets its excitation directly from the sense organ (figure 3). To have a nerve cell body at the surface, for instance in the skin, is not practical. Nerve cells die when the cell body is destroyed and do not regenerate. The cell body of a neuron can survive the loss of its fibre, however, and send out another process of cytoplasm to replace it. Hence, the cell body of the first neuron, in the chain that begins in a sense organ, protects itself by moving away from the surface. As it does so it leaves in its wake a strand of cytoplasm to retain its connection with the sense organ. In this way the cell body of the first neuron in such a chain acquires two nerve fibres, one that extends out to the sense organ and one that extends centrally to conduct the impulse to the cell body of the next neuron in the chain.

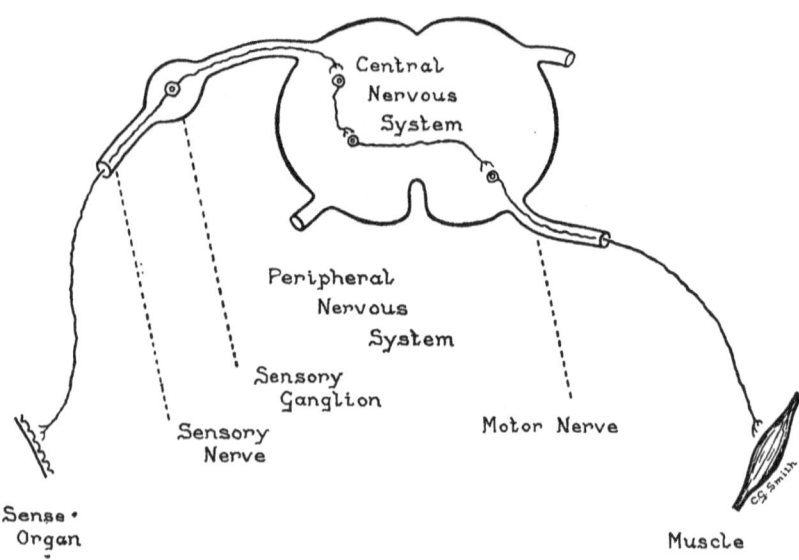

FIGURE 3. A chain of nerve cells extending from a sense organ to striated muscle. Note the parts of the chain that lie in the peripheral nervous system.

III. THE PARTS OF THE CHAIN OF NEURONS—LINKING A SENSE ORGAN WITH AN EFFECTOR—THAT ARE IN THE PERIPHERAL NERVOUS SYSTEM

The peripheral nervous system is made up of nerves plus swellings on those nerves called ganglia. The nerves are sensory, motor, or mixed. The sensory nerves carry impulses toward the central nervous system, the motor nerves carry impulses from the central nervous system to the muscles and glands. Each sensory nerve has a swelling near its attachment to the central nervous system. This is a **sensory ganglion**. Motor nerves have swellings too but these are found only on those motor nerves that supply smooth muscle, heart, and glands. These swellings are the **autonomic ganglia**. We will now examine these parts of the peripheral nervous system and locate the cell body and the nerve fibre of each of the successive neurons in the chain connecting a sense organ with an effector.

Figure 3 shows the location of the cell body and the nerve fibre or fibres of each of the neurons in the chain linking a sense organ with striated muscle. We see that all the neurons in such a chain have their cell bodies in the central nervous system except for neuron number one, which has its cell body in the sensory ganglion. Its nerve fibres, a distal one to the sense organ and a proximal one that enters the central nervous system, are the components of the sensory nerve. The nerve fibre of the last neuron in the functional unit is a constituent of the motor nerve. The cell body of the last neuron is, as already stated, inside the central nervous system along with the cell bodies and nerve fibres of all the intermediary neurons of the chain, regardless of their number. There may be many or few.

Figure 4 shows the location of the cell body and the nerve fibre or fibres of each of the neurons in the chain linking a sense organ with either smooth muscle, heart muscle, or gland. In this case all the neurons have their cell bodies in the central nervous system except two, that is neuron number one and the last neuron in the chain. The cell body of neuron one is in the sensory ganglion and its fibres are in the distal and proximal parts of the sensory nerve. The cell body of the last neuron is in a ganglion on a visceral motor nerve, that is, one of the autonomic ganglia. The nerve fibre of this last neuron in the chain is in the "postganglionic" motor nerve which extends from the ganglion to the effector. The "preganglionic" part of the motor nerve, the part that conducts impulses from the central nervous system to the auto-

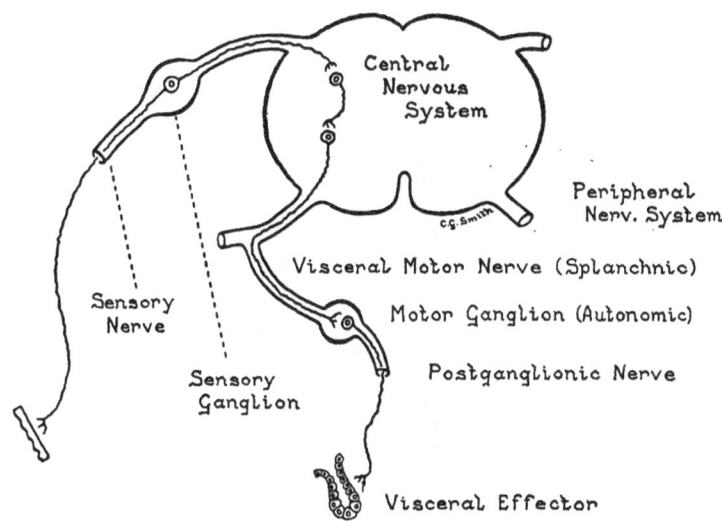

FIGURE 4. A chain of nerve cells extending from a sense organ to a gland, one of the visceral effectors. Note the parts of the chain that lie in the peripheral nervous system.

nomic ganglion, contains the nerve fibre of the second last neuron in the functional unit. The "preganglionic" motor fibre has grown out of the central nervous system in pursuit of the emigrant cell body of the terminal neuron. Note that there are synapses in the autonomic ganglia but there are no synapses in sensory ganglia.

IV. DEFINITIONS
A. A NEURAL PATHWAY

A neural pathway is a chain of nerve cells linked together end to end to convey excitation from one part of the body to another, for instance, from a sense organ to a muscle or from one part of the nervous system to another.

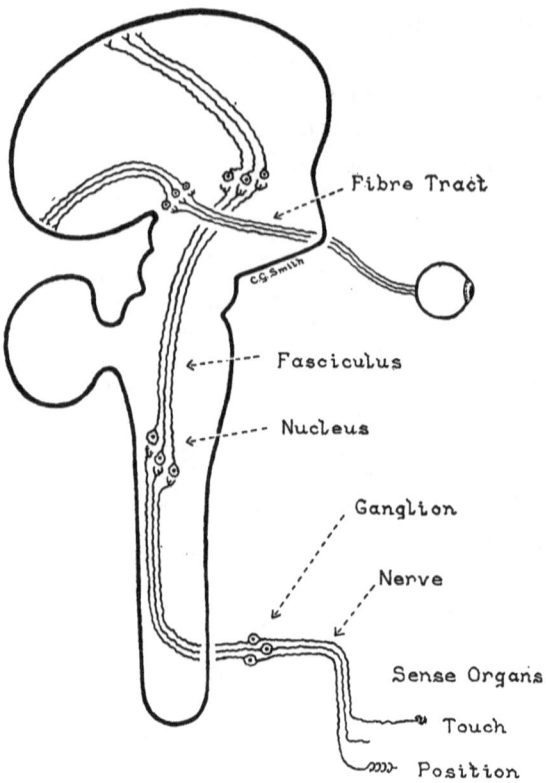

FIGURE 5. The components of a fibre tract, a fasciculus, a nucleus, a ganglion and a nerve.

B. A GANGLION AND A NUCLEUS

Neural pathways that are related functionally begin or end in the same part of the nervous system and they usually come together to form a more or less compact, cable-like structure. In such a cable-like structure the cell bodies of adjacent chains lie side by side and produce a local swelling because of their bulk. Such a swelling is called a ganglion if it is in the peripheral nervous system or a nucleus if it is in the central nervous system (figure 5).

C. A NERVE, A FASCICULUS, AND A FIBRE TRACT

A nerve is a cable-like bundle of nerve fibres in the peripheral nervous system. A fasciculus is a bundle of nerve fibres in the central nervous system, and a fibre tract is a cable-like bundle of fibres, all of which have a similar origin, termination, and function. Fibre tracts are found only in the central nervous system because there is no functional segregation of nerve fibres in mixed nerves.

D. A RELAY STATION OR SYNAPSE

A relay station or synapse is the part of a pathway where impulses are transferred from the end of the nerve fibre of one cell to the cell body of another. Hence relay stations are located in nuclei and also in all ganglia except the sensory ganglia. There are no synapses in sensory ganglia.

E. GREY MATTER AND WHITE MATTER

The central nervous system when cut across is seen to consist of two substances, one grey, the other white. The grey matter contains cell bodies of neurons, the white matter is made up of nerve fibres. The myelin sheaths give the fibre masses their whiteness. Grey matter is friable, white matter is tougher and when hardened, that is, coagulated by formaldehyde, it will split in the direction of its fibres like wood.

F. A COMMISSURE (A JOINING TOGETHER)

This is a band of grey or white matter that connects a portion of the central nervous system on the one side of the midline with its fellow of the opposite side.

G. A DECUSSATION

The crossing of the midline by an ascending or descending fibre bundle is called a decussation. Since corresponding bundles of the right and left sides of the central nervous system cross at the same place, they characteristically criss-cross through each other.

V. THE PARTS OF THE CENTRAL NERVOUS SYSTEM

The central nervous system is a tubular structure. Early in its development it has a uniform diameter and a relatively thin wall of uniform thickness. The differentiation of the central nervous system begins with the enlargement of the rostral end of the tube to form the brain. As the brain grows larger it is subdivided into a forebrain, a midbrain, and a hindbrain by two constrictions (figure 6). Secondary

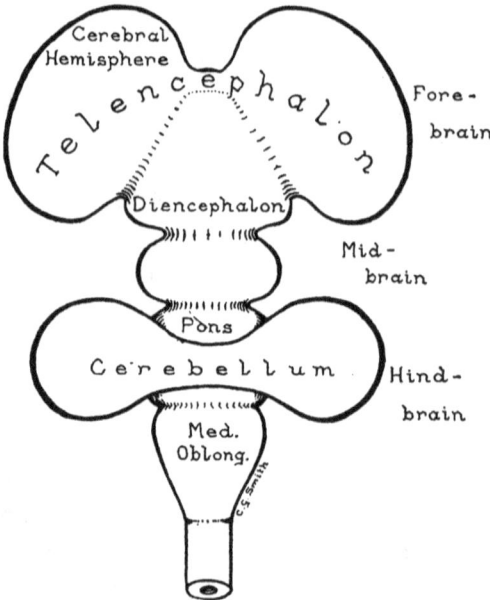

FIGURE 6. The subdivisions of the brain.

constrictions appear later to divide the forebrain and also the hindbrain into rostral and caudal segments.

The rostral part of the forebrain is called the telencephalon (end brain); the caudal part of the forebrain is called the diencephalon (between brain). Almost all of the telencephalon is evaginated to form the right and left cerebral hemispheres, which in the adult form the largest part of the brain. The portion of the telencephalon that

INTRODUCTION

remains in the midline does not enlarge and, in the mature brain, it is reduced to a thin membrane that stretches from one hemisphere to the other. Fibres that connect the right and left hemispheres cross the midline in this membrane.

The rostral part of the hindbrain is called the pons segment; the caudal part is the medulla oblongata. The cerebellum develops on the dorsal aspect of the pons segment. It takes the form of a dumbbell-like mass set across the back of the pons.

Figure 6 shows the parts of the brain as seen in a dorsal view. Figure 7 is a diagram showing the relationships of the parts of the

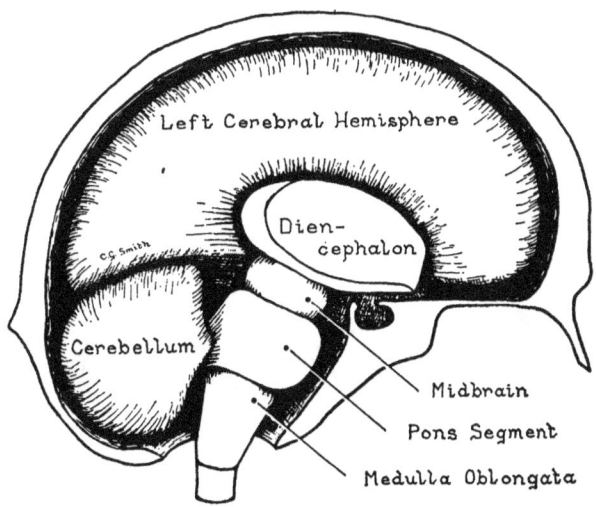

FIGURE 7. A midsagittal section of the cranium to show the position of each part of the brain. The right cerebral hemisphere was removed.

brain to the skull in a midsagittal section. These two diagrams show that the brain has an axial portion or stem consisting of medulla oblongata, pons segment, midbrain, and diencephalon. The medulla oblongata, pons, and midbrain are in line and rest against the sloping surface of the base of the skull formed by the basiocciput, basisphenoid, and dorsum sellae. The diencephalon is wedge-shaped with superior and inferior surfaces and a vertical anterior border. The posterior part of the inferior surface of the diencephalon is adherent to the superior surface of the midbrain. The anterior part of the diencephalon rests on the roof of the sella turcica.

VI. THE CAVITY OF THE CENTRAL NERVOUS SYSTEM

The cavity of the central nervous system has, at an early stage of development, a uniform diameter throughout its length. When the cerebral hemispheres develop the cavity acquires a right and a left pocket-like extension into each of them. The cerebellum develops as a thickening of the dorsal wall of the fourth ventricle and therefore has no central cavity. As growth and differentiation proceed, the walls of the tubular nervous system grow thick in some parts and remain thin in others. In the process, the central cavity is encroached on in some parts, but remains large in others. The larger, chamber-like parts are called ventricles, the small-bored passages that connect them are called canals.

Figure 8 shows diagrammatically the form of the central cavity in a midsagittal section of the brain. The cavity of the spinal cord is a

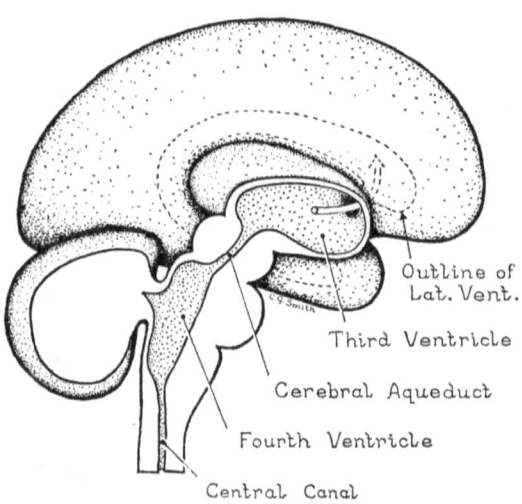

FIGURE 8. The parts of the central cavity of the brain and spinal cord as seen in a midsagittal section.

central canal about 1 mm. in diameter. This canal extends into the caudal half of the medulla oblongata and then enlarges to form the most caudal of the ventricles of the brain, the *fourth ventricle*. This is in the hindbrain. It has a thick ventral wall or floor. Its roof is formed by the cerebellum and two thin membranes, one rostral and one caudal to the cerebellum. These two membranes are (a) the anterior, rostral, or superior velum, and (b) the posterior, caudal, or inferior velum, respectively.

In the midbrain the cavity is again reduced to a canal about 2 mm. in diameter called the *cerebral aqueduct*. In the diencephalon the cavity enlarges to form the *third ventricle*. This is a median, cleft-like space, with a membranous dorsal wall, or roof, a thin rostral wall called the lamina terminalis (this is the midline part of the telencephalon), and a thin floor or ventral wall. In the lateral wall of the third ventricle, within the angle formed by the roof and the lamina terminalis, we find the opening of a short canal, the *interventricular foramen*. This canal drains the lateral ventricle, the cavity of the cerebral hemisphere. The *lateral ventricle* is a C-shaped tubular space. Its outline is illustrated in figure 8 as it would appear projected onto the medial surface of the hemisphere. The medial wall of the ventricle is thin, part of it is membranous; the lateral wall is very thick. The right and left lateral ventricles are numerically the first and second in the series of four ventricles of the brain.

VII. THE PRODUCTION, CIRCULATION, AND REMOVAL OF CEREBROSPINAL FLUID

The central nervous system is immersed in a watery fluid called cerebrospinal fluid (figure 9). Most of this is secreted into the cavity of the central nervous system by structures called choroid plexuses. One of these is in the medial wall of each lateral ventricle, one in the roof of the third ventricle, and one in the roof of the fourth ventricle. The fluid escapes from the cavity of the brain by way of three perforations that develop relatively late in foetal life in the part of the roof of the fourth ventricle located in the medulla oblongata. When the fluid reaches the external surface of the brain it fills the subarachnoid space and acts as a watery cushion for the brain and cord. The fluid is then transferred from the subarachnoid space to venous channels (dural sinuses) in the dura mater through villus-like projections of the outer surface of the arachnoid membrane. These may be large enough to be seen along the superior margin of the hemisphere after detaching the dura.

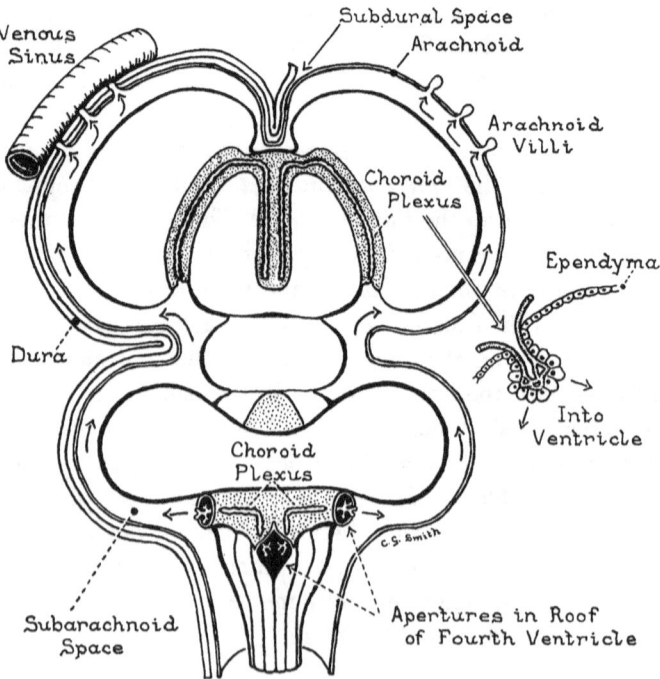

FIGURE 9. The circulation of the cerebrospinal fluid.

The choroid plexus, the chief source of the cerebrospinal fluid, has two components. One is the wall of the ventricle, which is reduced to paper thickness and consists almost entirely of cuboidal epithelium one cell thick. This is part of the ependyma which is the membrane lining the cavity of the central nervous system. The second component of a choroid plexus is a plexus of capillaries applied to the outer surface of the thin wall of the ventricle. The plexus presses against the membranous wall and causes it to bulge into the ventricle. The cuboidal cells of the wall of the ventricle are closely applied to the vessels of the plexus and take on the characteristics of gland cells. They extract the fluid from the capillaries on one side and secrete it into the ventricle on the other.

2. The Spinal Cord

I. FORM AND EXTERNAL FEATURES

THE SPINAL CORD is a cylindrical structure, flattened somewhat dorsoventrally so that its cross-section has an oval outline (figure 11). The maximum diameter does not exceed half an inch. At its rostral end it is continuous with the medulla oblongata at the margin of the foramen magnum (figure 10). The transition is gradual. At its caudal end it

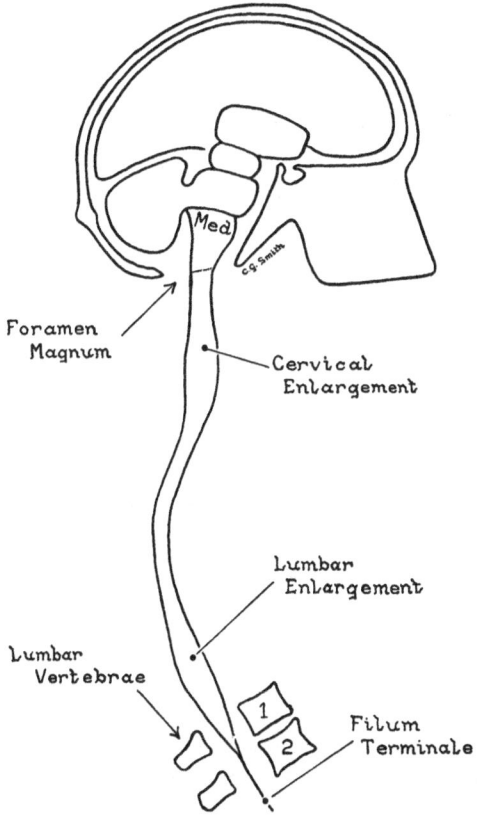

FIGURE 10. The form and the relationship of the spinal cord to the vertebral column.

tapers to a point at the level of the upper border of the second lumbar vertebra. A fine, thread-like strand of neuroglia overlaid by pia mater stretches from the caudal end of the cord to the back of the lowest coccygeal vertebra.

The spinal cord is segmented, one segment for each pair of spinal nerves, but there is nothing in the structure of the cord to betray this (figure 11). The cell masses and the fibre masses of adjacent segments

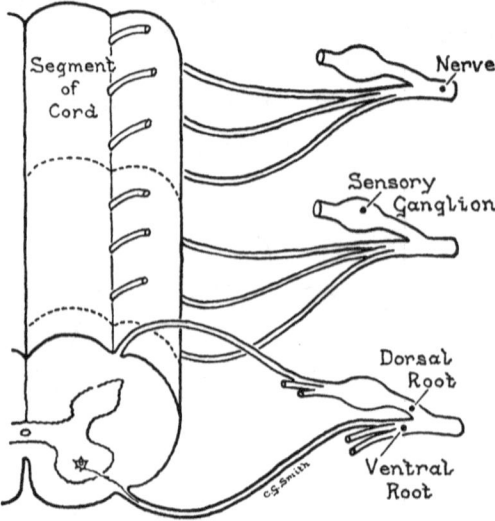

FIGURE 11. The attachment of a spinal nerve to its segment of the spinal cord.

join without a demarcation line and the diameters of a given segment do not vary throughout its length. The lack of a segmental enlargement of the cord, where a nerve joins it, is explained by the manner of attachment. Instead of entering or leaving the cord in one bundle, the nerve fibres of each nerve are first segregated into a sensory dorsal root and a motor ventral root, and then the fibres of each of these are evenly distributed in a series of filaments along the length of the segment. The line of attachment of the filaments of the dorsal root is marked by a shallow longitudinal groove located at the junction of the dorsal and the lateral surfaces. A similar, less distinct groove, located at the junction of the lateral and ventral surfaces marks the line of attachment of the filaments of the ventral roots. The filaments of adjacent segments form a continuous series and for this reason it is not possible to identify the segments after the roots are cut. Two other grooves extend the length of the cord; one in the midline dorsally is shallow, the other in the midline ventrally is very deep, about one-third the diameter of the cord. It contains a fold of pia mater and blood-vessels and is a feature used in identifying the ventral side of a cross-section.

In concluding this account of the external features, it is to be noted that a segment varies in size with the size of its nerve. This is the reason for the fusiform enlargements of the cord, one cervical and one lumbar. The cervical enlargement is made up of upper-limb segments C.5 to Th.1. The lumbar enlargement is made up of lower limb segments L.3 to S.3.

II. THE INTERNAL STRUCTURE OF THE CORD

The structure of the spinal cord is relatively simple and shows so little variation throughout its length that it may be difficult to recognize the level of a cross-section (figure 12).

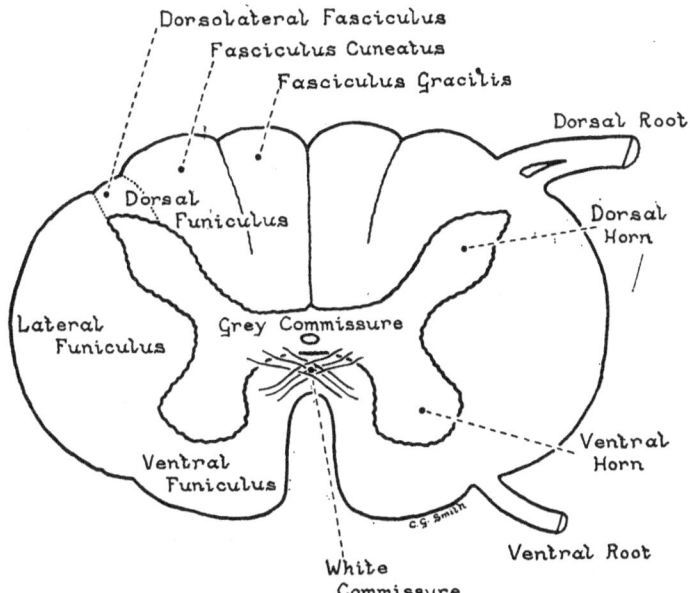

FIGURE 12. A cross-section of the spinal cord showing the subdivisions of the grey and white matter.

A. THE CENTRAL CAVITY OF THE CORD

The central cavity is 1 mm. in diameter and is located in the median plane close to the floor of the ventral median fissure. In the course of development the dorsal portion of the large cavity of the spinal cord is obliterated by the coming together of its lateral walls. Hence the central canal of the mature cord is the persisting ventral part of the original large cavity. Its obliteration is completed in some cases after the age of forty.

B. THE GREY MATTER

The grey matter surrounds the central canal as it does in the embryonic central nervous system but it loses the regular oval outline it has in the embryo. This is because cell proliferation during development is concentrated at two sites, one ventrolateral, opposite the site of exit of the motor nerve roots, and the other dorsolateral, opposite the site of the incoming dorsal root fibres. The cells that accumulate ventrolaterally are the cell bodies of the motor nerve fibres. These are very large and very numerous and together form a massive, horn-like extension called the ventral or **anterior horn**. The groups of cells in the ventral horn are the motor nuclei of the cord. The cell bodies that accumulate dorsolaterally serve as relay stations for the impulses arriving via the dorsal roots; they are the sensory nuclei of the cord and they form the **dorsal horn**. Thus the grey matter of a cross-section of the cord has the outline of a capital H. The cross-piece, called the **grey commissure**, contains the central canal and the nerve cells that surround it.

The cells of the grey matter of each segment of the cord are grouped to form nuclei (figure 13). Some of these are present in all segments, others in only a few. The nuclei in adjacent segments join to form an unbroken column of cells.

1. THE NUCLEI THAT ARE PRESENT IN ALL SEGMENTS
a) The Substantia Gelatinosa

This nucleus forms the tip of the dorsal horn. Its cells are small and there are very few myelinated fibres in it. This may account for its jelly-like consistency in the fresh state. Its cells serve as a local association mechanism for the dorsal horn.

b) THE CENTRAL NUCLEUS (PROPER SENSORY NUCLEUS)

This nucleus forms the bulk of the dorsal horn. Its cells are large or of medium size. They give rise to fibres that ascend to the brain.

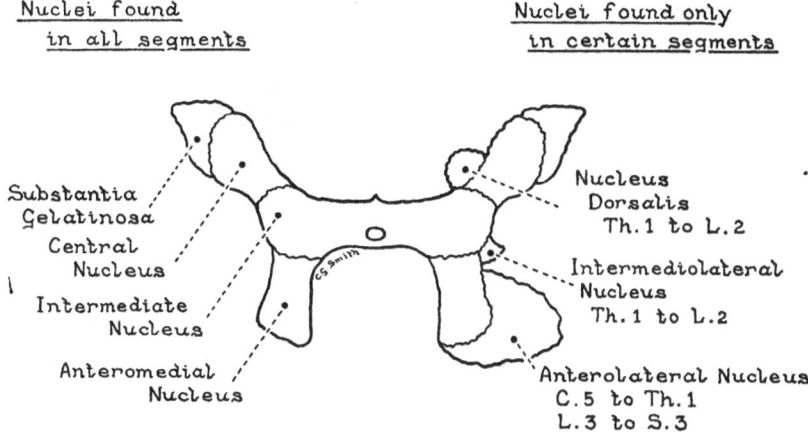

FIGURE 13. The nuclei in the grey matter of the spinal cord.

c) THE INTERMEDIATE NUCLEUS

This nucleus is between the dorsal and the ventral horns and it includes the cells of the grey commissure. Its cells are small or medium in size. They are interposed between the sensory and the motor nuclei as internuncial neurons.

d) THE ANTEROMEDIAL NUCLEUS

This forms the medial part of the ventral horn in the segments of the limb, and the whole of the ventral horn in other segments. Its cells are large and give rise to the motor fibres of the nerves to the trunk muscles.

2. THE NUCLEI FOUND ONLY IN PORTIONS OF THE CORD

a) THE ANTEROLATERAL NUCLEUS OF THE UPPER-LIMB SEGMENTS

This lies lateral to the anteromedial nucleus in the anterior (ventral) horn of segments C.5 to Th.1. Its cells are very large and give rise to the motor nerve fibres of the nerves to the muscles of the upper limb.

b) The Anterolateral Nucleus of the Lower-limb Segments

This lies lateral to the anteromedial nucleus in the anterior (ventral) horn of segments L.3 to S.3. Its cells are very large and give rise to the motor fibres of the nerves to the muscles of the lower limb.

c) The Nucleus Dorsalis (Clarke's Column)

This is located on the medial side of the central nucleus of the posterior (dorsal) horn in segments Th.1 to L.2. It serves as a relay station on a pathway to the cerebellum from sense organs located chiefly in muscles, tendons, and joint capsules.

d) The Intermediolateral Nucleus

This is a small nucleus, as seen in a cross-section, located lateral to the intermediate nucleus and projecting slightly into the lateral funiculus. It is present in segments Th.1 to L.2. Its cells are small and multipolar. They give rise to the preganglionic fibres of the sympathetic nervous system.

C. THE WHITE MATTER OF THE CORD

The bulk of the white matter of the cord is made up of fibres coursing longitudinally. The fibres that accumulate dorsal to the grey commissure completely fill the space between the two dorsal horns and are separated in the midline only by a neuroglial septum. Ventral to the grey commissure, however, the number of fibres is much smaller and a ventral fissure separates the fibres of the right and the left sides. The fibre bundle dorsal to the grey commissure and between the median septum and the dorsal horn is the **dorsal funiculus**. The bundle ventral to the grey commissure and between the ventral horn and the ventral fissure is the **ventral funiculus**. The rest of the fibres are in the **lateral funiculus**, a large bundle located lateral to the dorsal and ventral horns and separated from the dorsal and ventral funiculi by the fibres of the dorsal and ventral nerve roots.

The fibres of the dorsal funiculus are branches of sensory nerve fibres that are on their way to the brain. The number of these increases progressively at successively higher levels. The fibres that come in at lower levels of the cord are crowded medially, as they ascend, by fibres coming in at higher levels. In the cervical region, where the dorsal funiculus contains fibres from both the lower limb and the upper limb, a septum of neuroglia lies between them. This splits the dorsal funiculus into a medial bundle of lower-limb fibres called the **fasciculus gracilis** and a lateral bundle of upper-limb fibres called the **fasciculus cuneatus**.

The lateral and the ventral funiculi contain both ascending and descending fibres. These have their cell bodies in the spinal cord and in the brain respectively.

In addition to the longitudinal fibres that make up the funiculi, there are a few fibres that course transversely. If a dorsal root is traced into the cord from its attachment opposite the tip of the dorsal horn, the myelinated fibres will be seen to separate from the rest and form a bundle that courses medially, insinuating itself between the dorsal horn and the dorsal funiculus. This bundle gradually grows smaller as fibres turn into the dorsal horn. Branches of these fibres, as already stated, ascend in the dorsal funiculus. The unmyelinated fibres of the dorsal root form a lateral bundle that enters the gelatinous substance at the apex of the dorsal horn. Before doing so, however, the fibres of this bundle give off short branches that ascend between the apex of the dorsal horn and the surface of the cord; thus is formed a slender fasciculus of ascending fibres. It is included with the dorsal funiculus because its fibres are branches of sensory nerve fibres. The ascending branches of the myelinated fibres in the dorsal root form the fasciculi gracilis and cuneatus. The ascending branches of the unmyelinated fibres in the dorsal root form the *dorsolateral fasciculus* (also known as the tract of Lissauer). In addition to the transversely coursing fibres of the dorsal roots, we find transversely coursing fibres of ventral roots. These are delicate strands that converge toward the ventrolateral sulcus of the cord.

The only other bundle of fibres coursing in the plane of a cross-section is to be found between the floor of the ventral fissure and the grey commissure. It is made up of fibres that have their cell bodies in the dorsal horn and are crossing the midline. It is misleadingly named the **white commissure**. The fibres are not commissural: they are part of an ascending pathway, as we shall see.

D. DISTINGUISHING FEATURES OF THE CROSS-SECTION OF THE SPINAL CORD AT DIFFERENT LEVELS

The nuclei of the grey matter and the proportion of grey matter to white as seen in a cross-section serve as clues to the level of a cross-section.

In the thoracic segments the dorsal and the ventral horns are very slender and the nucleus dorsalis bulges into the dorsal funiculus. A distinct but less noticeable feature is the intermediolateral cell column that projects slightly into the lateral funiculus.

In the limb segments the ventral horns are massive, and a section of the cervical enlargement can be distinguished from a section of the lumbar enlargement by the septum in the dorsal funiculus that separates the fasciculus gracilis from the fasciculus cuneatus.

In the segments above those of the upper limb there is much more white matter than grey. In the segments caudal to those of the lower limb there is much more grey matter than white.

E. THE PATHWAYS OF THE SPINAL CORD

The primary pathway of the cord is the reflex pathway (see figure 14). This connects a sense organ of a spinal nerve with an effector of a spinal nerve. Each primary pathway has two or more off-shoots that

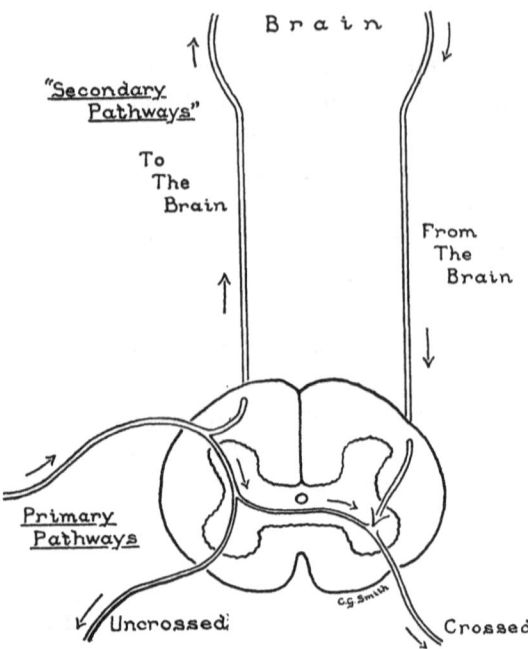

FIGURE 14. A diagram showing the relationship of the ascending and descending pathways to the primary (reflex) pathways.

ascend into the brain. These ascending pathways carry data from the sense organs to association regions in the brain. These association regions in turn are the origin of pathways that descend into the cord and discharge impulses into the terminal part of each primary pathway. Thus, the ascending and descending pathways make it possible for the brain to modify the response of an effector or, if it is not receiving impulses through a primary pathway, to activate it.

1. THE PRIMARY PATHWAYS

a) Segmental Pathways

These have a course in the cord that is limited to one segment. The pathways may enter and leave the segment on the same side or enter the segment on one side and leave it on the other.

The **uncrossed segmental pathway** enters the cord through the dorsal root of a spinal nerve, courses through the dorsal horn, the intermediate nucleus, and the ventral horn, and leaves the cord through the ventral root of the same spinal nerve. The pathway has at least two synapses, one in the dorsal horn and one in the ventral horn.

The **crossed segmental pathway** begins as a side branch of the uncrossed pathway that comes off as it passes through the intermediate nucleus. It crosses the midline in the grey commissure and courses through the intermediate nucleus and the ventral horn of the opposite side of the segment to leave the cord through its ventral root. This crossed pathway may have synapses in the intermediate nuclei of both sides and, without exception, in the ventral horn.

b) Intersegmental Pathways

These pathways connect the pathways of one segment with those of other segments, and make it possible for excitation entering the cord at one level to reach higher and lower levels. The intersegmental pathways are the fibres of cells that are found chiefly in the intermediate nucleus. These fibres ascend or descend in the nearest funiculus close to the grey matter to form the bundles known as the **fasciculi proprii** (*proprius*, one's own). The ends of these fibres turn into the intermediate nucleus of the higher and lower segments and synapse there with cells that convey the impulses to the ventral horn and thence to the effector.

22 BASIC NEUROANATOMY

2. THE PATHWAYS ASCENDING TO THE BRAIN

The pathways of the cord that ascend to the brain are fibres of cells of the dorsal horn or fibres of cells of the dorsal root ganglia. These ascending fibres end in different parts of the brain. Those having the same site of termination are grouped into cable-like bundles and these are crowded together on the outer surface of the grey matter to form all of the dorsal funiculus and about half of the lateral and the ventral funiculi. There is considerable intermingling of the fibres of adjacent bundles and in a normal section of an adult's cord they cannot be recognized. It is possible however, to recognize them in foetal specimens of the cord because the fibres of some bundles acquire their myelin sheaths earlier than others.

a) Sensory Pathways (Sensation)

(1) **Pain and Temperature**

The pain pathways and the temperature pathways course alongside one another and mingle to such an extent that the nuclei and fibre bundles of the two pathways cannot be distinguished (figure 15).

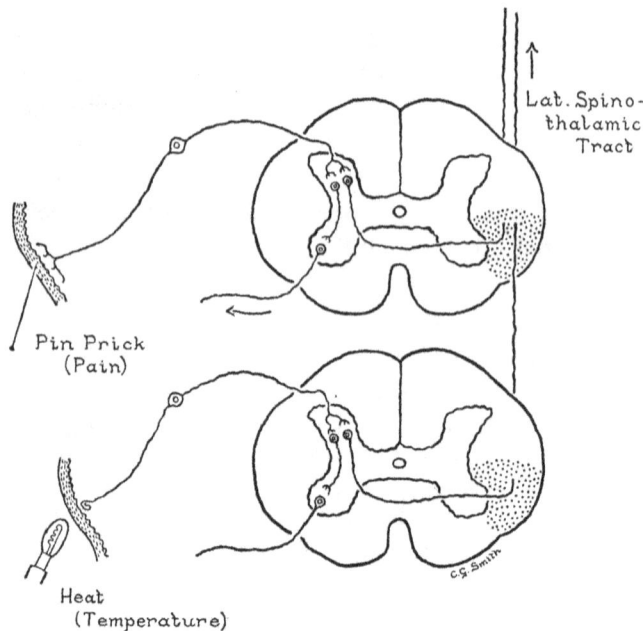

Figure 15. Pathways beginning in pain and in temperature sense organs.

SPINAL CORD

Each pathway begins as a branch of a sensory nerve fibre. This branch synapses with the cell located in the central nucleus which gives off the ascending fibre. These fibres cross the midline in the white commissure and enter the ventral (anterior) half of the lateral funiculus of the opposite side of the segment. There they ascend together in a bundle called the **lateral spinothalamic tract**. (This tract is so named because it is in the lateral funiculus and extends from the cord to the thalamus, a part of the diencephalon.) The fibres that join this tract at successively higher levels are added to the medial side of the bundle. Hence in cutting this tract for the relief of pain the cut must reach the grey matter to sever all the pathways that enter the cord below the level of the section.

(2) **Touch**

The impulses from a sense organ of touch ascend in two pathways, one on the same side of the cord, the other on the opposite side (figure 16). Hence an injury to one side of the cord will not affect the sense of touch appreciably.

(a) **The uncrossed touch pathway.** The ascending fibres of this

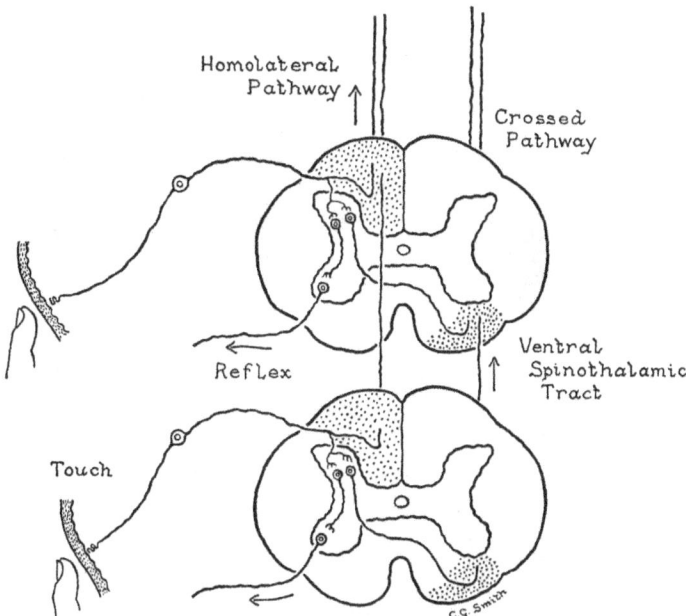

FIGURE 16. Pathways beginning in touch sense organs.

pathway are branches of the sensory nerve fibres that have their cell bodies in the dorsal root ganglion. These branches leave the sensory nerve fibres just before they enter the dorsal horn and they ascend in the dorsal funiculus. The fibres that enter the dorsal funiculus of each segment are crowded medially by those entering the segment immediately above; hence lower-limb fibres are medial to those from the upper limb in the cervical region of the cord.

(*b*) **The crossed touch pathway.** The long ascending fibres of this pathway come from cell bodies in the central nucleus of the dorsal horn. These cells are activated by the terminals of sensory nerve fibres that come from sense organs of touch. The ascending fibres cross the midline in the white commissure and enter the ventral (anterior) funiculus of the opposite side. There they ascend together in a bundle called the **ventral (anterior) spinothalamic tract.**

(3) **The Pathway of the Position Sense**

This ascending pathway carries impulses to the brain that excite awareness of the position of the parts of the body (figure 17). For

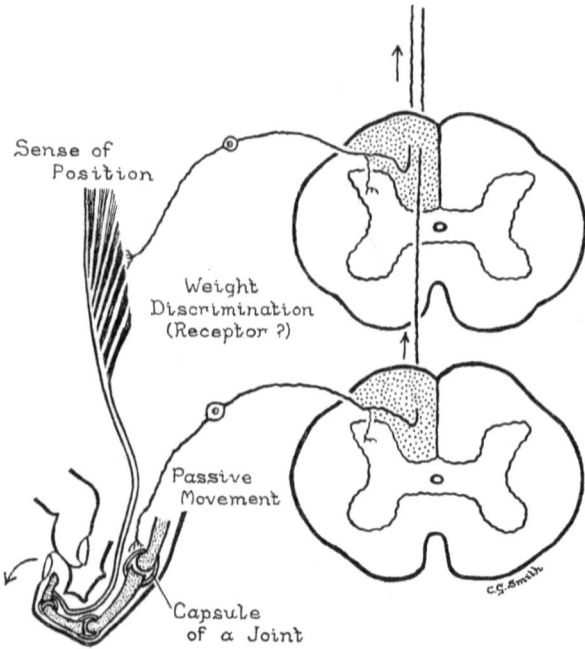

FIGURE 17. Pathways beginning in sense organs in muscles and in joint capsules. Interruption of the ascending pathway results in a loss of the sense of position.

instance, it tells us if the hand out of sight (behind the back) is open or closed. The impulses come from sense organs in muscles, tendons, and joint capsules, but chiefly from the joint capsules.

The fibres of this pathway are long branches of the fibres of sensory nerve fibres that have their cell bodies in the dorsal root ganglion. These branches leave the sensory nerve fibres just before they enter the dorsal horn. They ascend in the dorsal funiculus mingling with the fibres carrying impulses from sense organs of touch. Like these latter fibres they are crowded toward the midline by those that enter the dorsal funiculus in the segment immediately above.

b) PATHWAYS TO THE CEREBELLUM

These pathways convey data that are used by the cerebellum in the regulation of muscle contraction. Cutting these pathways does not impair sensation. It deprives the cerebellum of information it needs concerning the position of the parts of the body. This impairs its function and the result is awkwardness.

Several pathways reach the cerebellum from the cord. These probably have different functions but how they differ is not known. It is possible that they may carry impulses from different sense organs or from different parts of the body. At the moment we do not know enough about these pathways to make information concerning their individual characteristics of practical value.

The pathways to the cerebellum, like the sensory pathways, are either ascending branches of sensory nerve fibres or are secondary fibres from cells in the dorsal horn. Most of the secondary fibres, however, unlike the comparable fibres of the sensory pathways, remain on the same side of the spinal cord. Hence a unilateral lesion of the cord leads to a functional impairment (awkwardness) on the same side of the body.

(1) **The Ventral Spinocerebellar Pathway**

A branch of a dorsal root fibre enters the dorsal horn and synapses with a cell in the central nucleus (figure 18). Some of these cells send their fibres to the surface of the lateral funiculus (ventral half) of the same side. Others send their fibres to the same part of the lateral funiculus of the other side. The crossing is in the white commissure. The reason why some fibres cross in the cord is not known. There would appear to be no reason for it because they return to their own side within the substance of the cerebellum.

26 BASIC NEUROANATOMY

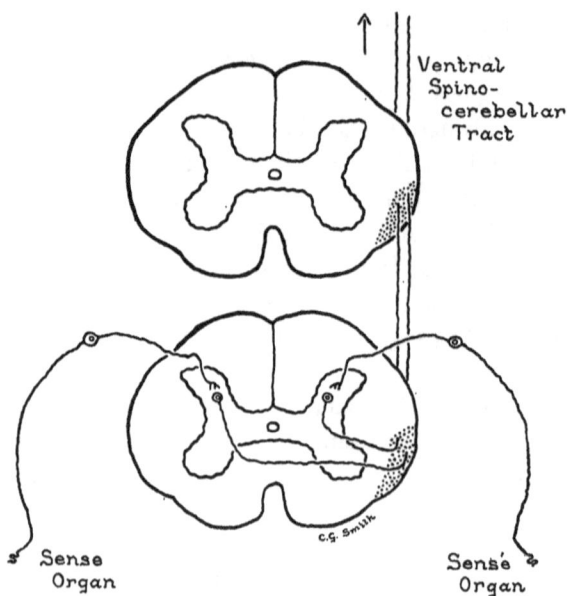

FIGURE 18. The ventral (anterior) spinocerebellar tract.

(2) **The Dorsal Spinocerebellar Pathway**

This pathway is described here as one that carries impulses from each of the nerves below the sixth thoracic (figure 19). Impulses that enter the cord above this level follow a different course to reach another relay station in the medulla oblongata. It is known that the dorsal root fibres entering the cord below the midthoracic region ascend in the dorsal funiculus through as many as eight segments before turning into the grey matter to synapse with a cell in the nucleus dorsalis. Hence, a branch of a dorsal root fibre ascending from the fifth lumbar segment would reach its relay station in or near the eleventh thoracic segment. Fibres that enter the cord at a higher level will ascend to correspondingly higher segments to reach their relay stations (figure 20). From the nucleus dorsalis fibres pass laterally to the surface of the dorsal half of the lateral funiculus on the same side and there ascend in a bundle called the dorsal spinocerebellar tract.[1]

(3) **The Pathway to the Cerebellum from the Upper Half of the Body**

The ascending fibres of this pathway are branches of the sensory

[1] R. E. Yoss, *Jour. Comp. Neur.*, LXLVII (1952), p. 5.

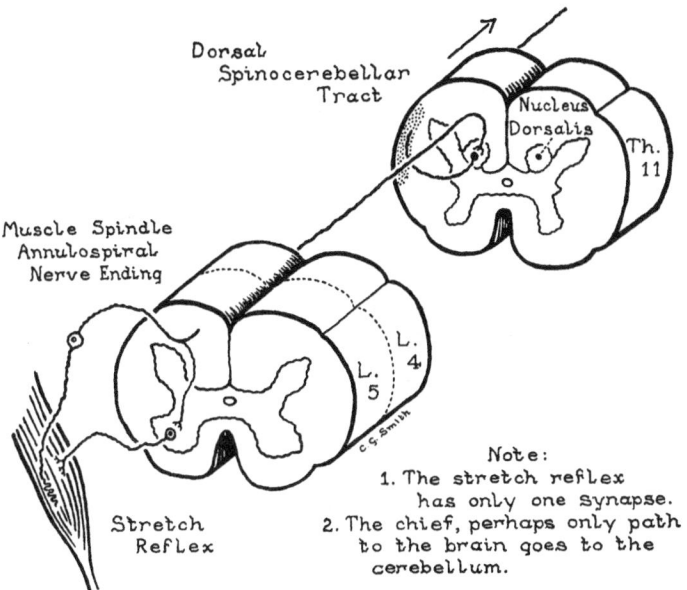

FIGURE 19. The dorsal (posterior) spinocerebellar tract.

nerve fibres in the upper half of the body. They come off the dorsal root fibres just before they enter the dorsal horn and ascend in the lateral part of the dorsal funiculus. They, like the fibres that end in the nucleus dorsalis, ascend a long way to reach their relay station. In this case it is in the brain, just above the cord. It is called the accessory cuneate nucleus and its fibres like those of the nucleus dorsalis join the dorsal spinocerebellar tract to ascend into the cerebellum (figure 20).

3. THE PATHWAYS DESCENDING FROM THE BRAIN

Pathways descend from the brain to feed impulses into the terminal part of the primary pathways (figure 14). They end in the intermediate nucleus or in a ventral horn nucleus. These pathways either initiate voluntary movement or they reflexly adjust posture and thus play an essential role in the execution of all voluntary movement. The fibres of the descending pathways that initiate voluntary movement are in two compact bundles, the lateral and ventral corticospinal tracts. The other descending fibres are dispersed within the lateral and the ventral (anterior) funiculi and will be described with the nuclei of the brain from which they come.

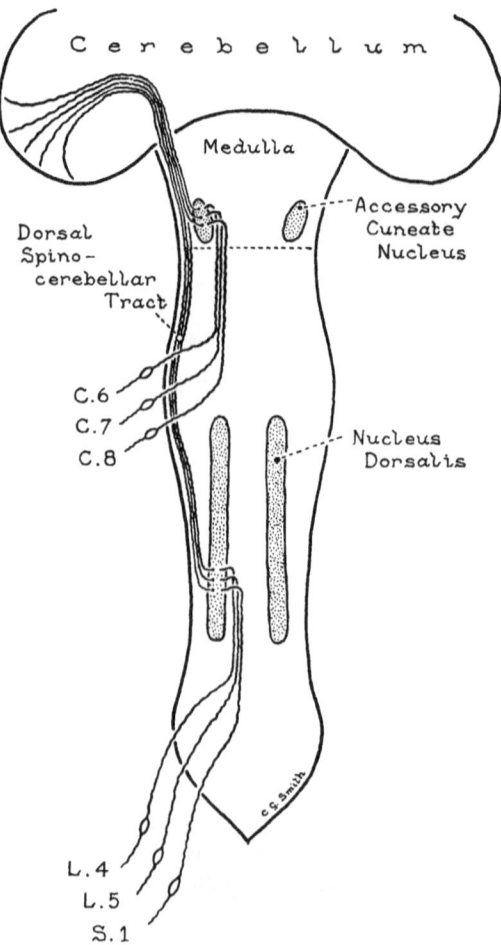

FIGURE 20. The dorsal spinocerebellar pathway and the pathway corresponding to it from the upper limb.

a) THE LATERAL CORTICOSPINAL TRACT

Most of the fibres of this tract come from the opposite side of the brain (figure 21). It is located in the dorsal half of the lateral funiculus, sandwiched between the dorsal horn and the dorsal spinocerebellar tract. Some of its fibres end in each segment of the cord. They synapse with cells of the intermediate nucleus or with the cells of the motor nuclei in the ventral horn. Impulses discharged in the

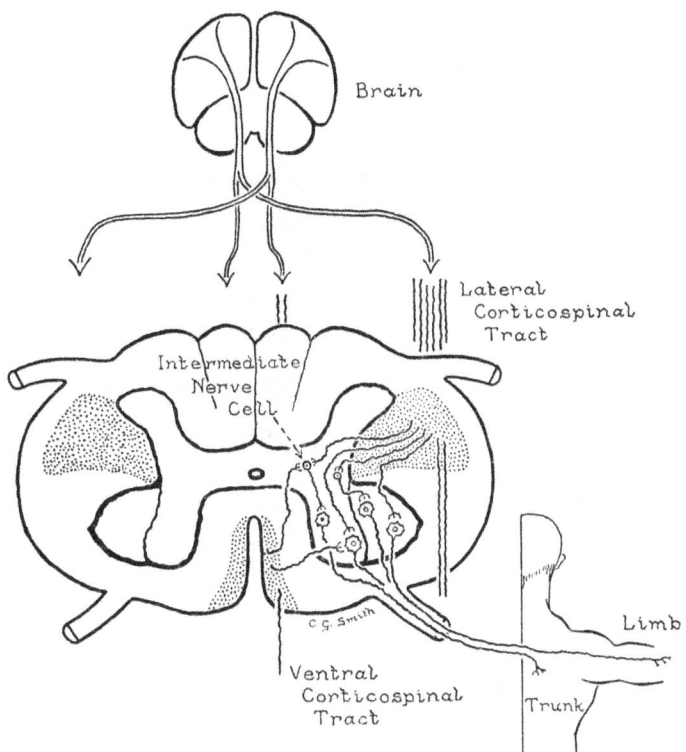

FIGURE 21. The corticospinal tracts.

intermediate nucleus are relayed to the ventral horn nucleus of the same segment.

b) THE VENTRAL CORTICOSPINAL TRACT

This is a very small bundle in the ventral funiculus that varies in size and may be absent (figure 21). All its fibres come from cells in the brain on the same side as the tract. If the tract is unusually small, or absent, its missing fibres are included in the lateral corticospinal tract as homolateral fibres of that bundle. Some of its fibres end in the anteromedial nucleus of each segment. Since this nucleus also receives crossed fibres from the lateral corticospinal tract it can be activated by either the right or the left side of the brain. This means that if the pathway from one side of the brain is cut, the muscles supplied by this nucleus—the trunk muscles—will not be paralyzed.

3. The Medulla Oblongata

I. FORM AND DIMENSIONS

THE MEDULLA OBLONGATA may be likened to an inverted four-sided pyramid that has its apex cut off for the attachment of the spinal cord (figure 22). It is about 30 mm. long and at its larger upper end it is about 25 mm. wide and 15 mm. thick (dorsoventral diameter).

II. EXTERNAL FEATURES

A. THE VENTRAL (ANTERIOR) SURFACE

The ventral surface has just one feature, the pyramid (figure 23). This is a tapered longitudinal ridge, hence its name. It is separated

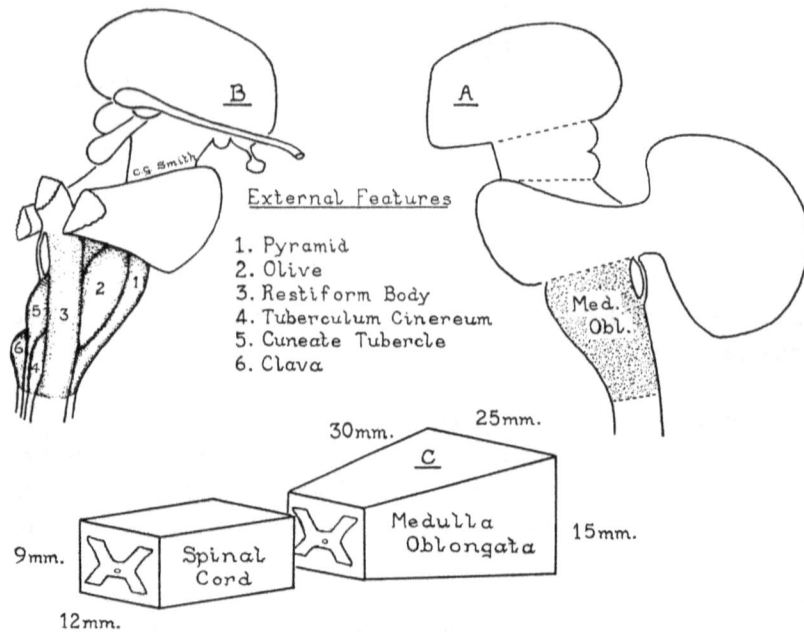

FIGURE 22. A: A schematic outline of the lateral aspect of the brain stem showing its subdivisions. B: The lateral aspect of the brain stem showing the external features of the medulla. C: The dimensions of the medulla oblongata represented as a basal portion of a four-sided pyramid.

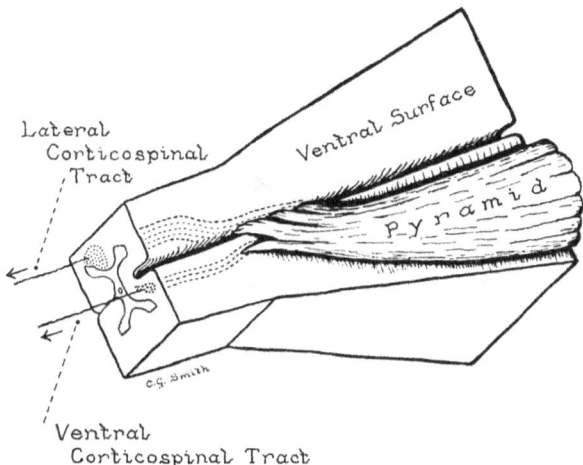

FIGURE 23. The ventral aspect of the medulla and the first cervical segment showing the course of the fibres of the pyramid.

from its fellow of the opposite side by a ventromedian fissure and at its narrower, caudal end it appears to continue into the ventral funiculus of the cord. The pyramid is partly continuous with the ventral funiculus because it is made up of corticospinal fibres and some of these descend in the ventral corticospinal tract. The rest of the fibres of the pyramid leave the ventral surface as they enter the cord and cross the midline to form the core of the dorsal half of the lateral funiculus. These are the fibres of the lateral corticospinal tract. They may fill the ventromedian fissure as they cross.

B. THE DORSAL (POSTERIOR) SURFACE

The dorsal surface has three features (figures 24, 25). Two of these are longitudinal ridges formed by the fasciculus gracilis and the fasciculus cuneatus, which ascend onto the caudal half of the back of the medulla. The third feature is a triangular membrane that forms the roof of the caudal part of the fourth ventricle. The two ridges enlarge before tapering to a point and form club-shaped swellings, the clava (fasciculus gracilis) and the cuneate tuberce (fasciculus cuneatus). The terminal increase in size of these fasciculi is due to a relay station in each. These relay stations are the nuclei gracilis and cuneatus. The cells of these nuclei send their fibres ventrally and across the midline to ascend to the diencephalon and therefore the clava and cuneate tubercle both taper to a point.

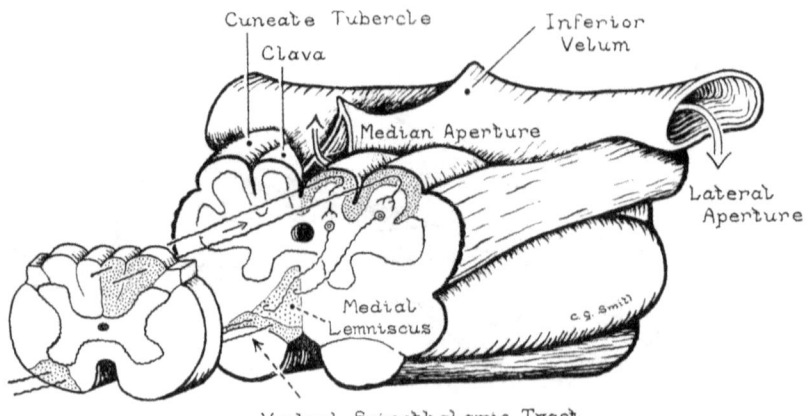

FIGURE 24. The dorsolateral aspect of the medulla with a segment removed to show the decussation of the sensory pathways located in the dorsal funiculus. The features of the dorsal aspect are labelled.

The departure of the sensory pathways from the dorsal part of the tubular central nervous system permits the central cavity to enlarge toward the dorsal surface to form the chamber-like fourth ventricle. The ependymal lining of the fourth ventricle, plus the supporting tissue left behind in the dorsal part of the tube by the departing sensory pathways, form a membrane. This is the roof of the ventricle in the medulla and in the human brain it may be called the inferior velum. It is triangular with its apex located a little caudal to the centre of the dorsal surface of the medulla. The caudal half of its lateral border is attached to the clava and above this to the cuneate tubercle. This progressive replacement of the clava and then the cuneate tubercle by a membrane (the inferior velum) indicates that the sensory pathways leave the surface in order, first the most medial ones and then the successively more lateral ones.

The inferior velum contains the choroid plexus of the fourth ventricle. This is a T-shaped structure. The stem of the T is in the midline and the cross-piece extends the full width of the upper part of the velum. The inferior velum also has perforations which allow the cerebrospinal fluid to escape from the ventricular system. There are three of these, one at each lateral angle and one at the inferior angle in the midline. Each of the lateral openings is at the end of a tubular outpouching of the fourth ventricle known as the lateral recess of the fourth ventricle. This extends ventrally around the side of the medulla.

C. THE LATERAL SURFACE

One nucleus and two fibre bundles appear on this surface (figure 25). One of the fibre bundles extends the length of the medulla. It

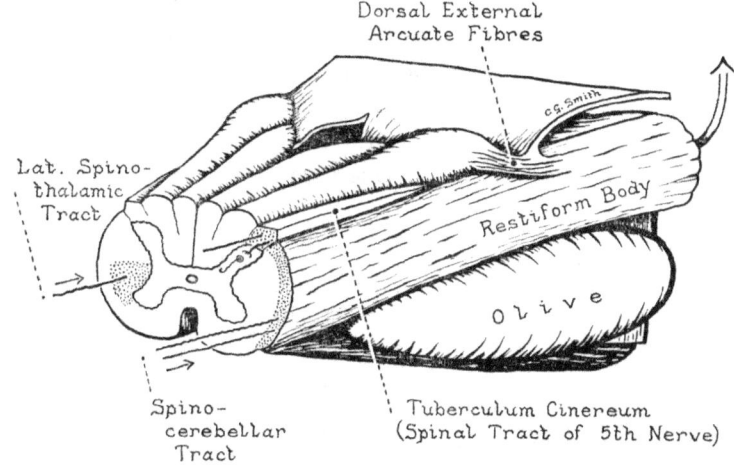

FIGURE 25. The dorsolateral aspect of the medulla. The features of the lateral aspect are labelled.

covers almost all the lateral surface of the medulla where the latter joins the cord and here its fibres are those of the spinocerebellar tracts. These fibres shift dorsally as they ascend toward the stalk of the cerebellum and permit an ovoid mass called the olive to bulge to the surface ventral to them. The spinocerebellar fibres are joined by other cerebellar fibres from inside the medulla and also by fibres from the cuneate tubercle. The latter are the dorsal external arcuate fibres. They come from the accessory cuneate nucleus and carry impulses to the cerebellum from the nerves of the upper half of the body. Thus the spinocerebellar tracts are joined by other tracts to the cerebellum to form a thick, rope-like bundle, the restiform body (*restis*, rope).

The olive which crowds the spinocerebellar tracts dorsally is an elongated, cigar-shaped nucleus covered laterally by a thin layer of fibres. It intervenes between the restiform body and the pyramid and extends from the upper end of the medulla to within 8 mm. of the cord.

The third feature of the lateral surface is a narrow, low ridge about 1 mm. wide that lies along the dorsal border of the spinocerebellar fibres intervening between them and the cuneate tubercle. It begins near the upper end of the cuneate tubercle coming to the surface just caudal to the dorsal external arcuate fibres which cross it obliquely. It extends down onto the upper two segments of the cord lying in the groove where the dorsal roots enter the cord. This indicates that it lies along the crest of the dorsal horn. It is a bundle of fine, unmyelinated fibres and therefore the ridge has an ashen grey colour and is known as the tuberculum cinereum (*cinereus*, ashy).

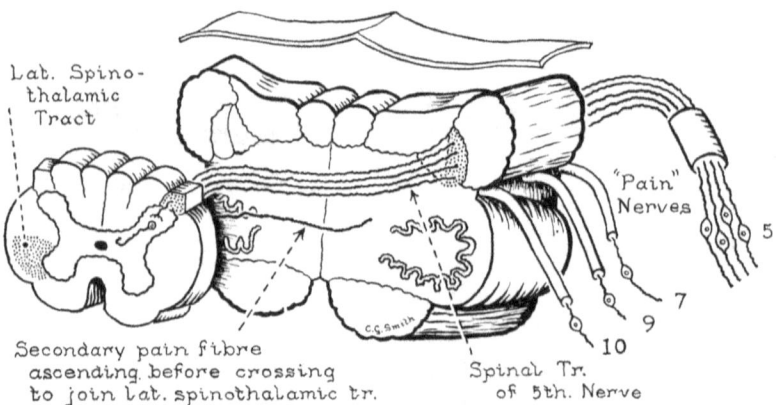

FIGURE 26. The dorsolateral aspect of the upper portion of the medulla and the first cervical segment to show the course of the fibres of the cranial nerves that carry impulses from sense organs excited by painful stimuli and changes in temperature.

The fibres of this bundle enter the brain in the fifth, seventh, ninth, and tenth cranial nerves (figure 26). They have their cell bodies in the sensory ganglia of these nerves and come from pain and temperature sense organs. Most of these fibres belong to the fifth nerve, hence the bundle is known as the spinal (descending) tract of the fifth nerve. The fibres are descending to end in a long, slender nucleus that lies along the crest of the dorsal horn in the upper two cervical segments and extends up into the lower part of the medulla.

III. THE INTERNAL STRUCTURE

The change in internal structure between the lower and the upper end of the medulla is a gradual one but the medulla may be divided into three portions each with a characteristic feature. Beginning caudally where the medulla joins the cord there is a segment in which the fibres of the pyramid cross the midline. This is the region of the motor decussation. Above the level of the motor decussation, the sensory pathways of the dorsal funiculus cross the midline. This is the region of the sensory decussation. The upper portion of the medulla is known as the open medulla because it contains the fourth ventricle which has a very thin roof that is easily torn away.

A. THE REGION OF THE MOTOR DECUSSATION

This segment of the medulla is about 9 mm. long (figure 27). Its internal structure is like that of the cord except that the fibres of the pyramid are present on the front of the section and some of them are coursing across the midline to reach and help form the dorsal half of the lateral funiculus. These are the fibres of the lateral corticospinal tract. The crossing fibres pass through and obscure the fibres of the ventral funiculus. They also pass through the grey matter and partially isolate the ventral horn. The fibres of the pyramid that do not cross the midline help to form the ventral funiculus.

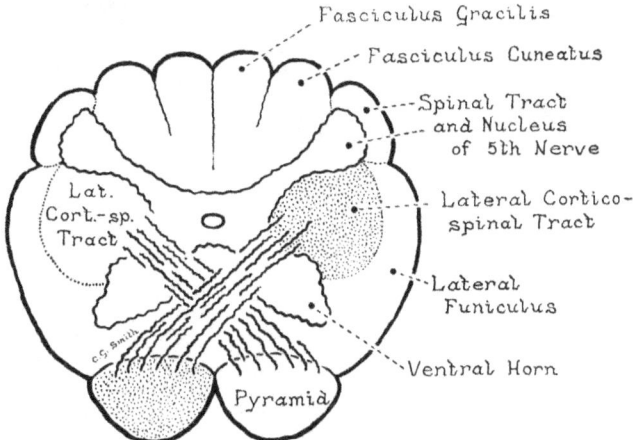

FIGURE 27. A cross-section of the medulla at the level of the decussation of the fibres of the pyramid.

B. THE REGION OF THE SENSORY DECUSSATION

The grey and white matter in sections of this region of the medulla (figure 28) can be related to those of the cord if the transition from cord to medulla is studied in serial sections. This reveals that the grey matter of the dorsal horn of the cord is prolonged into this region but rotates around the central canal until its crest points laterally. This rotation may be interpreted as a shift of the grey matter to fill the space which at lower levels is occupied by the lateral corticospinal tract. It also becomes evident in a study of serial sections that the dorsal horn is the source of the two nuclei that project dorsally into the core of the fasciculus gracilis and fasciculus cuneatus respectively. This is to be expected because the fibres of the fasciculus gracilis and cuneatus are ascending sensory nerve fibres. A third large nucleus in the dorsolateral part of the grey matter of these sections corresponds to the crest of the dorsal horn of the cord. This fittingly receives sensory nerve fibres also, but these sensory fibres are coming from cranial nerves, chiefly the fifth cranial nerve. They descend in the spinal tract of the fifth nerve, described with the features of the lateral

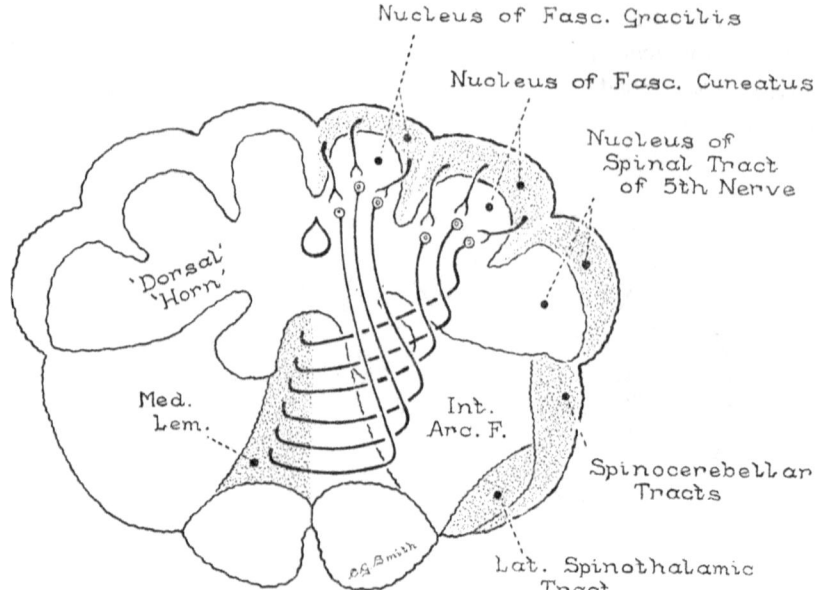

FIGURE 28. A cross-section of the medulla at the level of the decussation of the sensory pathways that make up the fasciculi gracilis and cuneatus.

surface. The tract forms a cap for this nucleus which is known as the nucleus of the spinal tract of the fifth nerve.

The grey matter corresponding to the ventral horn of the cord is relatively very small and poorly defined because the fibres of the cells of the nucleus gracilis and cuneatus are coursing ventrally through it before crossing the midline. These are the internal arcuate fibres. After they cross the midline they ascend in the region that corresponds to that of the ventral funiculus of the cord and they join there the ventral spinothalamic tract. Thus in this portion of the medulla, the uncrossed touch pathway of the cord joins the crossed pathway of the cord to form one bundle of fibres. This bundle is ribbon-like and located in the sagittal plane adjacent to the median plane, hence it is known as the medial lemniscus (*lemniscus,* ribbon).

C. THE LEVEL OF THE OPEN MEDULLA

The characteristics of a section at this level are the fourth ventricle and the inferior olivary nucleus which bulges laterally to form the olive (figure 29).

The fourth ventricle is a cleft-like space with a ventral and a dorsal wall. Its lower end is at the middle of the medulla where it is

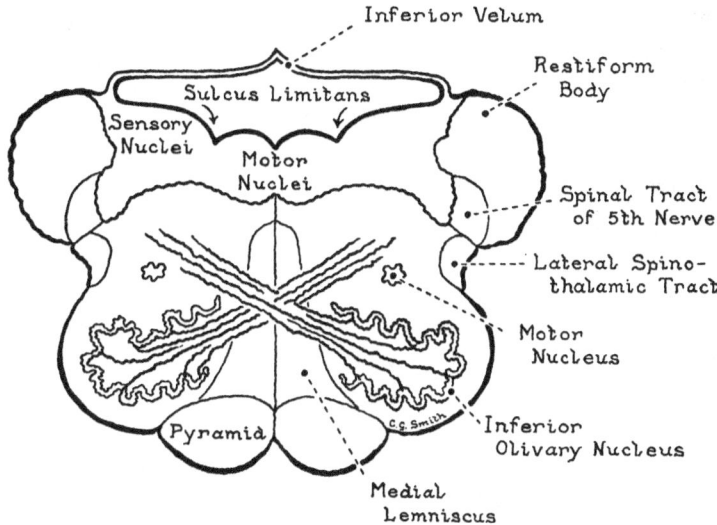

FIGURE 29. A cross-section of the open medulla. Note the decussation of the olivocerebellar fibres occurs at this level.

continuous with the canal which is prolonged upwards from the cord. The canal enlarges to form the ventricle as the medial and then the lateral sensory pathways depart from the dorsal wall of the neural tube. Therefore the ventricle attains its full size (width) above the sensory decussation. The width of the ventricle is still further increased at its upper, widest part by a tubular outpouching of the lateral angle of the inferior velum. This tube has the lateral aperture at its extremity. It extends laterally, crosses the back of the restiform body and adheres to it (figure 116). Hence the ventricle acquires a lateral recess. The choroid plexus is prolonged laterally in the roof of this recess and it usually projects into the subarachnoid space through the lateral aperture.

The inferior olivary nucleus, as seen in a transverse section, is a sac-like nucleus with its opening toward the median plane. The thin layer of nerve cells that form the wall of the sac is wrinkled as if to provide a maximal amount of surface. Its afferent fibres—chiefly from the higher levels of the brain—reach its outer surface and form a thin capsule for it. The efferent fibres of this nucleus all go to the cerebellum on the opposite side. They leave from its inner surface and thence through the opening of the sac they cross the midline and join the restiform body of the opposite side.

The other structures in the transverse section of the upper part of the medulla are cell masses and fibre bundles that have been encountered at lower levels. The grey matter in the floor of the ventricle has cell masses in its medial portion that correspond to the nuclei of the ventral horn of the cord. The cell masses in the lateral part correspond to the nuclei of the dorsal horn. The floor of the ventricle has a longitudinal sulcus, the sulcus limitans, which divides the floor into a sensory lateral portion and a motor medial portion. Although all the sensory nuclei are in the floor of the ventricle, all the motor nuclei are not. Some of the grey matter of the ventral horn does not shift dorsally and medially to the floor of the ventricle but continues on into the core of the medulla in line with the ventral horn of the cord. This is a very slender strand of cells, only two or three cells in a microscopic section, hence it does not appear as a grey mass in a freshly cut brain.

The core of each half of the medulla oblongata is the reticular substance. It is composed of widely separated cells that lie in a loosely woven network of fibres (*reticulum,* a web). This core surrounds the slender detached column of motor nuclei and is itself surrounded by grey matter and fibre tracts. Dorsal to it is the grey matter of the floor of the ventricle. Medial to it is the medial lemniscus which is applied to the median plane with its dorsal border close to the motor nuclei and its ventral border in contact with the pyramid. Ventral to the reticular substance is the inferior olive and lateral to it is the lateral spinothalamic tract which is crowded dorsally into a compact bundle by the inferior olivary nucleus. Dorsal to the lateral spinothalamic tract is the spinal tract of the fifth nerve which is covered on its lateral side by the restiform body.

4. The Pons Segment

I. FORM AND GENERAL CHARACTERISTICS

THE PONS SEGMENT (figure 30) is the portion of the hindbrain on which the cerebellum develops. It is situated between the medulla oblongata and the midbrain. Before the cerebellum develops the segment has a relatively simple form, somewhat like that of a cube. After the cerebellum develops, a large pathway comes down from the cerebral hemisphere and covers its ventral surface and most of

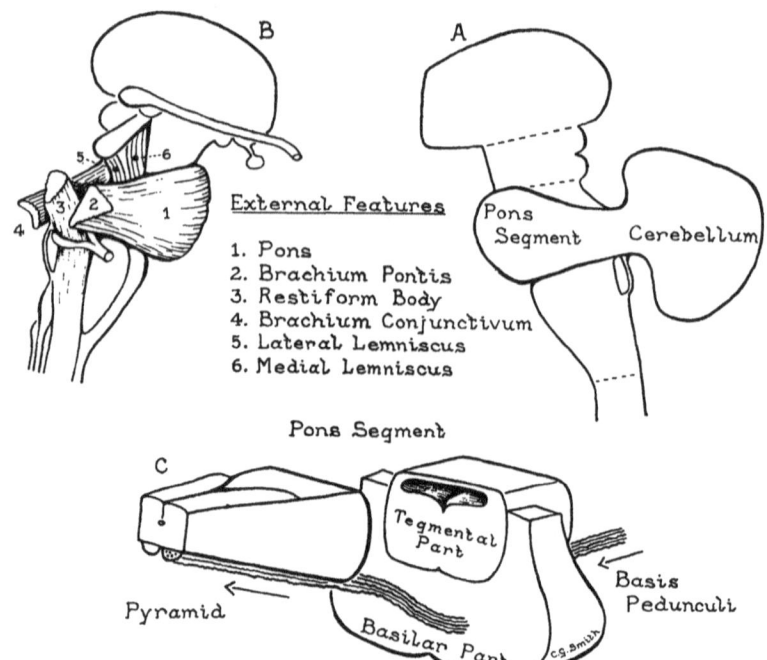

FIGURE 30. A: A schematic outline of the lateral aspect of the brain stem showing its subdivisions. B: The lateral aspect of the brain stem showing the external features of the pons segment. C: The pons segment isolated to show its tegmental and basilar portions.

its lateral surface. These superficial pathways—a right and a left—decussate on the front of the hindbrain and form a bridge-like transverse band called the pons which gives this subdivision its name. It will help us to understand the structure of the pons segment if we examine the steps in the development of the cerebellum and its afferent pathways.

Early in development, the cavity of the hindbrain has a membranous roof that forms not only the future inferior velum of the medulla, but also the whole of the dorsal surface of the undifferentiated pons segment (figure 31). At the lateral border of this membrane, near the middle of the hindbrain, the sensory nerve fibres of the internal

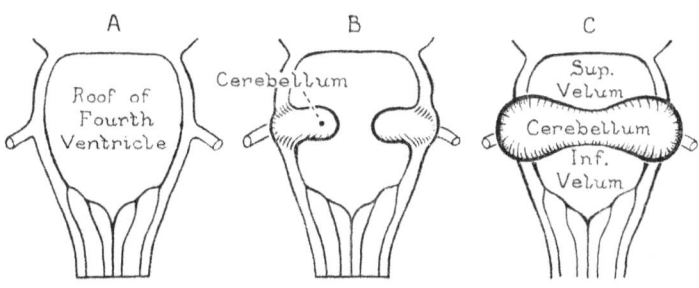

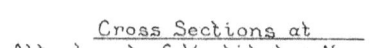

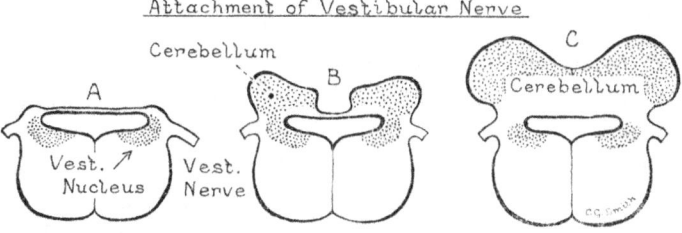

FIGURE 31. Three stages in the development of the cerebellum.

ear enter the brain. These are the fibres of the vestibular nerve. As they enter the brain, nerve cells accumulate around them to serve as a relay station. The cells of this relay station are in several groups and one of these enlarges to form the cerebellum.

The cells of the cerebellar group form a thin, superficial layer of grey matter called *cortex* (bark). As this grey layer enlarges it extends medially in the roof of the ventricle as a tongue-like projection. The right and left portions of the developing cerebellum meet and unite at the midline to form a bar of grey matter that divides the membranous roof of the ventricle into upper and lower portions, the *superior* and the *inferior vela*. The name velum (a sail) is apt because each of the membranes arches up into the stalk of the fully developed cerebellum like a sail. Later, as the cerebellum enlarges its functional scope, it receives afferent fibres that grow into it from sense organs in muscles. These are the fibres that form the restiform body which enters the lateral extremity of the cerebellum along with the vestibular nerve fibres.

The accumulation of fibres in the cerebellum elevates the cortex to form a ridge. This ridge grows higher and its crest grows wider but its attachment to the roof of the ventricle does not enlarge. Hence in a lateral view (profile) the cerebellum has the form of a bulb with a short stalk. The lateral portion of the cerebellum grows larger than the median part, and the cerebellum acquires the shape of a dumbbell. Meanwhile, the number of fibres in the stalk of the cerebellum increases and crowds out the cells that form the grey band connecting the cortex and the vestibular nucleus.

The third stage in the enlargement of the cerebellum begins with the arrival of a pathway from the cerebral hemisphere (figure 32). This pathway is a chain of two neurons. The first neuron has its cell body in the cerebral hemisphere and during the fourth month of foetal life the first contingent of fibres from these neurons grows caudally along the ventral surface of the brain stem as far as the front of the upper end of the hindbrain. There these fibres, called corticopontine fibres, meet a group of cells that are to relay impulses from them to the cerebellum. These cells come from a germinal centre in the dorsolateral part of the hindbrain and reach this site by migrating on the surface. They form the first portion of the developing pontine nucleus. After establishing themselves on the front of the pons segment each of the cells sends a nerve fibre medially and superficially across the midline to the lateral side of the pons segment where it and its fellows turn dorsally forming a compact bundle that enters the cerebellum lateral to the restiform body. This bundle is the brachium pontis. The crossing fibres from the right and left pontine nuclei criss-cross through each other at the midline and thus is formed

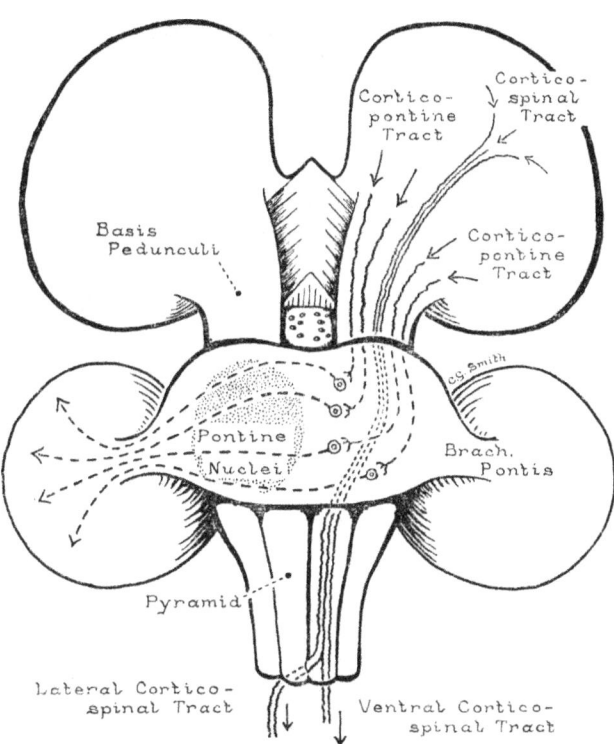

FIGURE 32. A diagram of the ventral aspect of the brain to show how the pons is formed by the pathways descending from the cerebral hemisphere to the cerebellum. To simplify the diagram only the pathways from the left hemisphere are charted.

a superficial, strap-like structure that extends across the front of the pons segment from the right extremity of the cerebellum to the left. This is the pons. It grows thicker and wider as successive contingents of corticopontine fibres and layers of pontine cells are added to its surface.

The pons is, phylogenetically, a new part of the brain. It is not present in reptiles or birds and when it does appear in the mammalian brain its size varies with the development of the cerebral hemisphere. Being a recently acquired structure it has been added to the outer aspect of the brain. In man it has become so large that it constricts the portion of the hindbrain that it embraces and reduces it to a slender isthmus uniting the medulla and the midbrain. This slender portion of the original hindbrain is called the **tegmentum** of the pons segment; the pons itself is called the **basilar part** of the pons segment (figure 30).

II. EXTERNAL FEATURES

A. DORSAL SURFACE

Figures 33 and 34 show the features of the dorsal surface as revealed when the cerebellum is detached. Let us begin with the area to which the cerebellum is attached. We note first that it extends across the caudal one-fourth of the dorsal surface. In its lateral part is the section

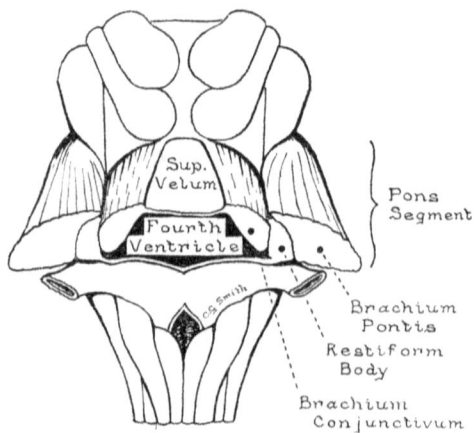

FIGURE 33. The dorsal aspect of the pons segment. The cerebellum has been removed.

of the brachium pontis, which brings impulses from the hemisphere. Medial to this is the section of the restiform body which brings impulses from the muscles, joints, and internal ear, and medial to this again is a section of a third bundle, the brachium conjunctivum, which is the efferent fibre bundle of the cerebellum. This efferent bundle, although it carries impulses destined for muscles of the trunk and limbs, does not pass caudally as it enters the brain stem. It courses up toward the midbrain along the lateral border of the superior velum gradually encroaching on it until the right and left brachia meet at the midline. Thus the superior velum, like the inferior, is triangular. The bases of these two membranes form the middle part of the stalk of the cerebellum. Between them is the dorsal recess of the fourth ventricle.

B. THE VENTRAL SURFACE

The ventral surface is covered by the portion of the pons that contains the right and left pontine nuclei. These produce right and left enlargements of the strap-like band of fibres and account for a median groove in which the basilar artery lies when it is not sinuous.

C. THE LATERAL SURFACE

The lateral portions of the pons become much narrower as they pass dorsally because the pontine nuclei are restricted to its ventral part. The fibres of the pons converge toward the attachment of the cerebellum and leave a small, triangular portion of the lateral surface uncovered. This exposes a broad ribbon of fibres that is passing upwards and dorsally crossing superficial to the brachium conjunctivum to reach the surface of the midbrain (figure 34). This ribbon

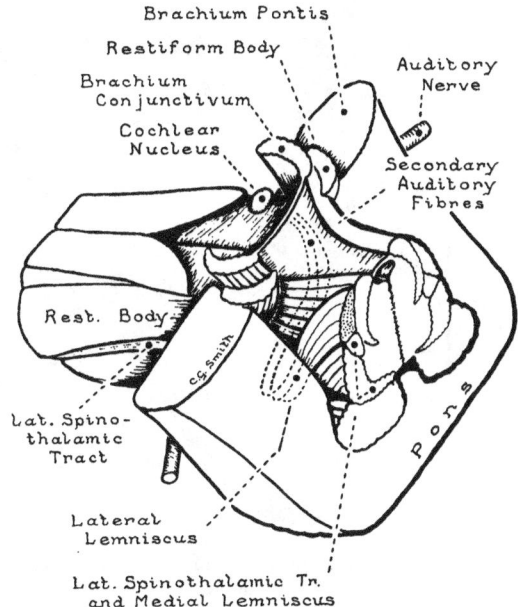

FIGURE 34. The dorsolateral aspect of the hindbrain. The cerebellum has been removed and the bundles of fibres in its stalk have been isolated. The auditory pathway from the left auditory nerve to the right lateral lemniscus is charted.

—about 8 mm. wide—is formed by two smaller ribbons each about 4 mm. wide. They cannot be distinguished except by tracing them to where they diverge. The dorsal one is the lateral lemniscus, the ventral the medial lemniscus. The course of these two pathways within the pons segment will be described with its internal structure.

III. THE INTERNAL STRUCTURE

A. THE TEGMENTAL PART

A section of this part of the pons segment resembles that of the medulla (less the pyramid) because both are portions of the embryonic hindbrain (figure 35).

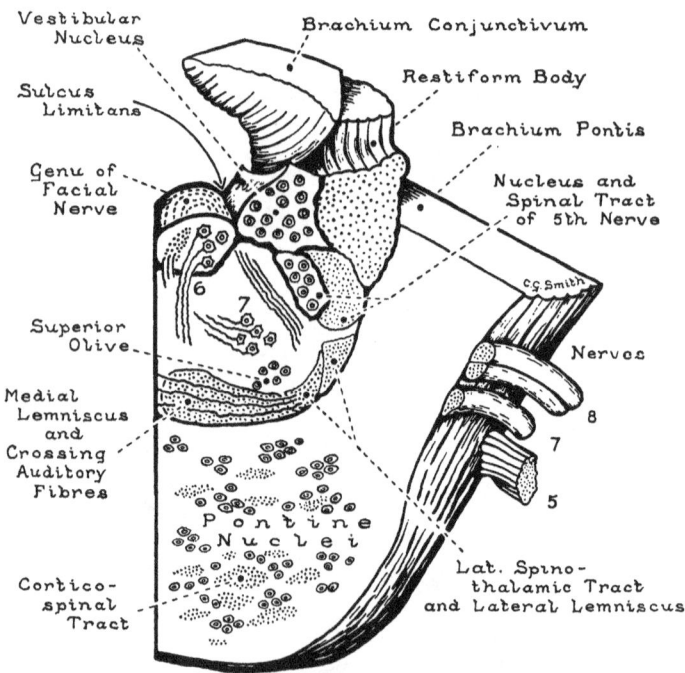

FIGURE 35. The caudal aspect of a thick slice of the pons segment in a plane near the medulla.

The floor of the fourth ventricle has a covering of grey matter with a sulcus limitans that divides it into medial and lateral portions. The medial portion contains motor nuclei, the lateral portion sensory nuclei.

In the centre of each half of the section there is, as in the medulla, a section of another motor nucleus (labelled "7" in figure 35). Ventrolateral to this nucleus, in the region corresponding to the site of the inferior olive, there is a small nucleus called the superior olive. It is in the caudal part of the pons segment in the plane of the attachment of the auditory nerve and is a relay station of this pathway.

The tegmentum has a core of reticular substance and the fibre tracts cover its outer aspect. In the medulla the medial lemniscus intervenes between it and the midline but in the pons this ribbon has glided onto the ventral side of the reticular substance and forms the outer border of the tegmentum. It continues to shift laterally as it ascends through the pons and we have seen that it appears on the lateral side of the pons segment just above the brachium pontis. The lateral spinothalamic tract (pain and temperature) is in the same location as in the medulla. It is lateral to the reticular substance, just dorsal to the superior olive (figure 35). Dorsal to this is the spinal tract of the fifth nerve and dorsal to this again is the restiform body which is turning dorsally to enter the cerebellum.

In addition to the tracts (pathways) listed above which are present in the medulla there is a new sensory pathway to be seen. This is the auditory pathway. It comes from sense organs in the cochlea located in the temporal bone. The sensory nerve fibres have their cell bodies in a ganglion in the cochlea and proceed to the brain in the acoustic nerve (eighth cranial nerve). The auditory fibres of this nerve reach the ventral side of the restiform body just caudal to the brachium pontis and pass medially across the restiform body in the floor of the lateral recess of the fourth ventricle. Here, in the floor of the lateral recess, the fibres reach their relay station, the **sensory cochlear nucleus** (figure 68). In the gross, it looks like a swollen terminal part of the nerve (figure 34). Most of the fibres of these cells cross the midline to ascend on the opposite side of the brain in a ribbon-like bundle called the lateral lemniscus. These fibres pass upward toward the midbrain as they cross the midline and therefore only short portions of them are seen in a cross-section. As they leave the cochlear nuclei some pass dorsal to the restiform body, others ventral to it and penetrate the reticular substance in scattered bundles. These converge as they course transversely and penetrate the medial lemniscus of the opposite side. After passing through and around the medial lemniscus they turn and pass upward mingling with the fibres of the lateral spinothalamic tract. This part of the auditory pathway is so named because it is a flattened, ribbon-like band located lateral to the medial lemniscus. The two lemnisci lie side by side and their fibres mingle to form one wide ribbon that can be seen on the lateral surface of the pons segment above the brachium pontis.

A small number of cochlear fibres do not cross the midline. They end in the superior olive which is a relay station on an uncrossed auditory pathway. The fibres of these cells join the crossed fibres in the adjacent lateral lemniscus.[1]

B. THE BASILAR PART

In a section of the pons segment at the level of the attachment of the cerebellum, the basilar part is a U-shaped mass that embraces the tegmentum. In its ventral part (anterior) the cells of the pontine nuclei are in layers separated by fibres coursing transversely to form the brachium pontis. The pontine nuclei also contain cross-sections of scattered bundles of fibres. Some of the fibres in these bundles are corticopontine, the rest are corticospinal. The latter fibres grow down from the hemisphere with the corticopontine fibres in successive contingents during the fourth month of foetal life. Each contingent of fibres is covered in turn by the arrival of a layer of cells. Hence both the corticospinal fibres and the corticopontine are contained in the flattened bundles that are embedded in the pontine nuclei. The bundles of corticospinal fibres come together toward the caudal part of the pontine nuclei and emerge in one bundle, the pyramid.

[1] A paper on the auditory pathway by Barnes, Magoun, and Ranson in the *Jour. of Comp. Neurol.*, LXXIX (1943), p. 129 may be consulted.

5. The Midbrain

I. FORM AND GENERAL CHARACTERISTICS

THE MIDBRAIN is the middle segment of the embryonic brain, hence its name. It does not acquire appendages and therefore remains small and retains a relatively uncomplicated form. Most of the ascending and descending pathways pass through it without interruption but it contains large relay stations that function as correlation centres in the control of postural movements. Of these the one that controls the movement of the head and eyes is on the dorsal surface in an elevation known as the superior colliculus. It is a characteristic feature of the midbrain in all vertebrates.

II. EXTERNAL FEATURES

The midbrain may be likened to a cube. It has four free surfaces, a dorsal, a ventral, and a right and a left free surface. Its inferior surface is attached to the pons segment and its superior surface is attached to the diencephalon. The features of its free surfaces are:

A. DORSAL SURFACE
Superior colliculus
Inferior colliculus

B. LATERAL SURFACE
Brachium of the superior colliculus
Brachium of the inferior colliculus
Medial lemniscus
Lateral lemniscus

C. VENTRAL SURFACE
Basis pedunculi (crus cerebri)
Interpeduncular fossa

The above features are illustrated in the diagram of figure 36C where they are represented as the coverings of a cube-shaped, isolated midbrain. All the features except for the interpeduncular fossa on the ventral surface are visible in a lateral view (figure 36B). In describing the external features we will use the medial lemniscus of the lateral surface as a key structure and work dorsally and ventrally from it.

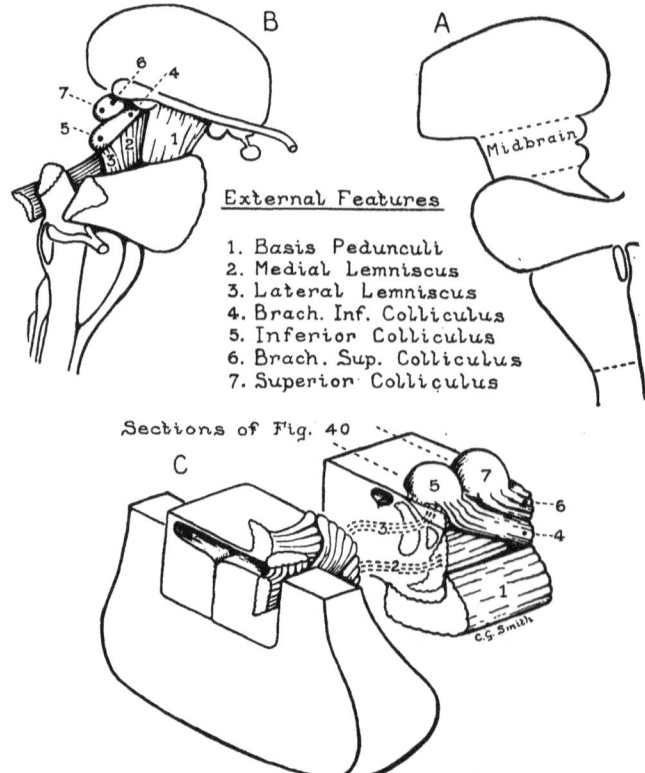

FIGURE 36. A: A schematic outline of the lateral aspect of the brain stem showing its subdivisions. B: The lateral aspect of the brain stem showing the external features of the midbrain. C: The midbrain showing its relationships to the pons segment.

THE MEDIAL LEMNISCUS

The medial lemniscus made up of fibres of the pathways for the senses of touch and position reaches the lateral surface of the brain as it ascends through the pons segment and it continues upward in this position onto the middle of the lateral surface of the midbrain. At the dorsal border of this ribbon-like bundle we find the lateral lemniscus (auditory pathway) and at its ventral border we find the large bundle of descending pathways called the basis pedunculi (crus cerebri). As the medial lemniscus ascends through the midbrain it is overlapped more and more by these pathways and as it reaches the upper end of the midbrain it is completely covered by them.

THE BASIS PEDUNCULI (CRUS CEREBRI)

The basis pedunculi, a rope-like bundle 15 mm. wide, contains the corticopontine and corticospinal tracts. It courses along the ventral surface of the midbrain separated from its fellow of the opposite side by a deep, wide fossa, the **interpeduncular fossa**. As the right and the left basis pedunculi are followed from the pons toward the diencephalon they diverge from the midline and come to lie more and more on the lateral surface, where they help to cover the medial lemniscus, as already pointed out.

THE LATERAL LEMNISCUS

The lateral lemniscus located at the dorsal border of the medial lemniscus makes a sharp turn onto the dorsal surface as it enters the midbrain to end in a relay station called the nucleus of the inferior colliculus. This cell mass bulges onto the dorsal surface to form a large, round eminence called the **inferior colliculus** (little hill). This covers the caudal half of the dorsal surface of the midbrain. Most of the neurons of this relay station are a part of the chain of neurons that serve as the auditory pathway to the cerebral hemisphere. Their fibres form a compact bundle, the **brachium (arm) of the inferior colliculus**, that extends upwards on the lateral surface. This bundle ascends superficial to the medial lemniscus and shifts ventrally as it does so until it contacts the dorsal border of the basis pedunculi at the upper end of the midbrain.

The small, triangular area of the lateral surface, dorsal to the brachium of the inferior colliculus, is covered by a band of fibres that takes the form of a ridge parallel to the brachium of the inferior colliculus. Its caudal end turns onto the dorsal surface to end in a dome-like eminence of grey matter that covers the rostral half of the dorsal surface. This is the **superior colliculus** and the fibre bundle that ends in it is its brachium, the **brachium of the superior colliculus**.

The pathways that make up the brachium of the superior colliculus are not the same in all mammals. In the rat the brachium of the superior colliculus is made up of optic nerve fibres (figure 37). These

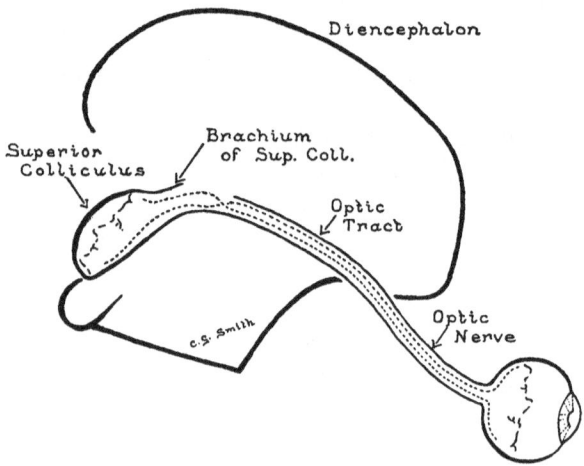

FIGURE 37. A diagram showing the composition of the brachium of the superior colliculus in a rodent.

come from the ventral surface of the diencephalon where the optic nerves are attached. When they enter the superior colliculus they spread out and end on its surface to provide a point-to-point projection of the retina. Thus the surface of the superior colliculus may be likened to the viewing screen of a television camera. The camera in this case is the eye. This viewing screen is connected with the muscles that move the eye and through these it automatically adjusts the eyes to maintain the picture of the moment and bring it into focus.

In the human brain the superior colliculus has lost its role as a viewing screen but retains its role as a controlling centre for automatic movements of the head and eyes. The viewing screen of the superior colliculus has been replaced by a much larger and improved screen, called the visual sensory area. This is located on the surface of the cerebral hemisphere (figure 38). Hence the optic pathway now bypasses the midbrain and goes directly from the diencephalon into the cerebral hemisphere. It is surprising, therefore, that a brachium

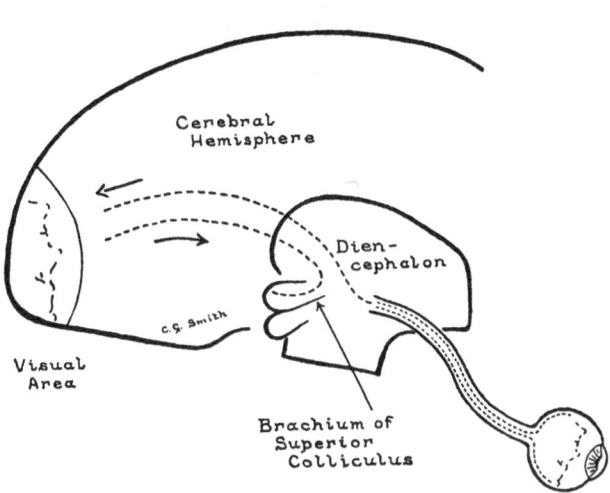

FIGURE 38. A diagram showing the composition of the brachium of the superior colliculus in man.

of the superior colliculus is present in the human brain. It is present and even looks like the brachium in the brains of lower animals although it is not made up of optic nerve fibres. The fibres it contains have their cell bodies in the cerebral hemisphere. These fibres come from the visual area and course first into the diencephalon and thence to the superior colliculus to form a newly constituted brachium. The fibres connect the visual area with the centre controlling the movements of the eyes. By means of this pathway the eyes automatically follow a moving object.

III. INTERNAL STRUCTURE

The midbrain has been likened to a stalk for the forebrain (figure 39). The forebrain, shaped like a half-sphere and made up of the two cerebral hemispheres and the diencephalon, is called the **cerebrum**. Hence the midbrain, on which this hemispherical mass rests, is called the **cerebral peduncle**. This explains why the large fibre bundle on its ventral or basal surface is called the basis pedunculi (basal part of the stem) or crus cerebri (leg of the cerebrum). The rest of the substance of the midbrain is divided arbitrarily into the **tectum** and

tegmentum. The tectum is dorsal to a transverse plane through the aqueduct; the tegmentum is ventral to it (figure 39).

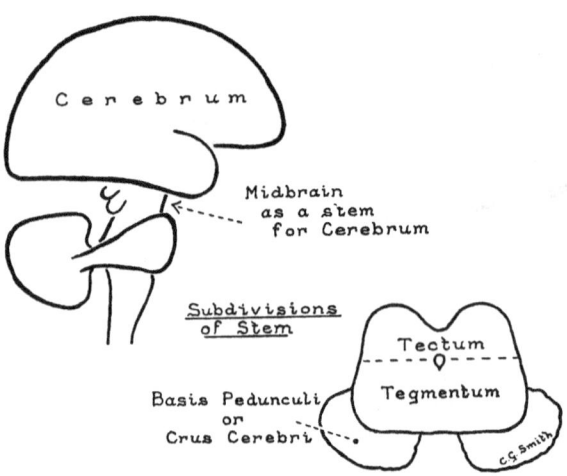

FIGURE 39. The midbrain as a cerebral peduncle (stem).

A. THE CAVITY OF THE MIDBRAIN, THE CEREBRAL AQUEDUCT

The cavity of the midbrain is reduced to a canal called the cerebral aqueduct (figure 40). It is in the median plane close to the dorsal surface. The name is fitting because it serves as a duct or drain for the cavities of the cerebrum which are filled with the watery cerebrospinal fluid. It is so small (its diameter is 2 mm. or less) that it is liable to be blocked by debris or closed by pressure from without.

B. GREY MATTER OF THE SECTION OF THE MIDBRAIN

1. THE CENTRAL GREY MATTER

This is the thick layer of grey matter that surrounds the aqueduct (figure 40). In its ventral part we find the three motor nuclei of the midbrain and the one sensory nucleus. These will be described with the cranial nerves but we may note here that just as in the hindbrain the motor nuclei are medial to the sensory nucleus and are to be found next to the median plane. The rest of the cells in the central grey matter are not grouped into well defined nuclei and their connections have not been established.

MIDBRAIN

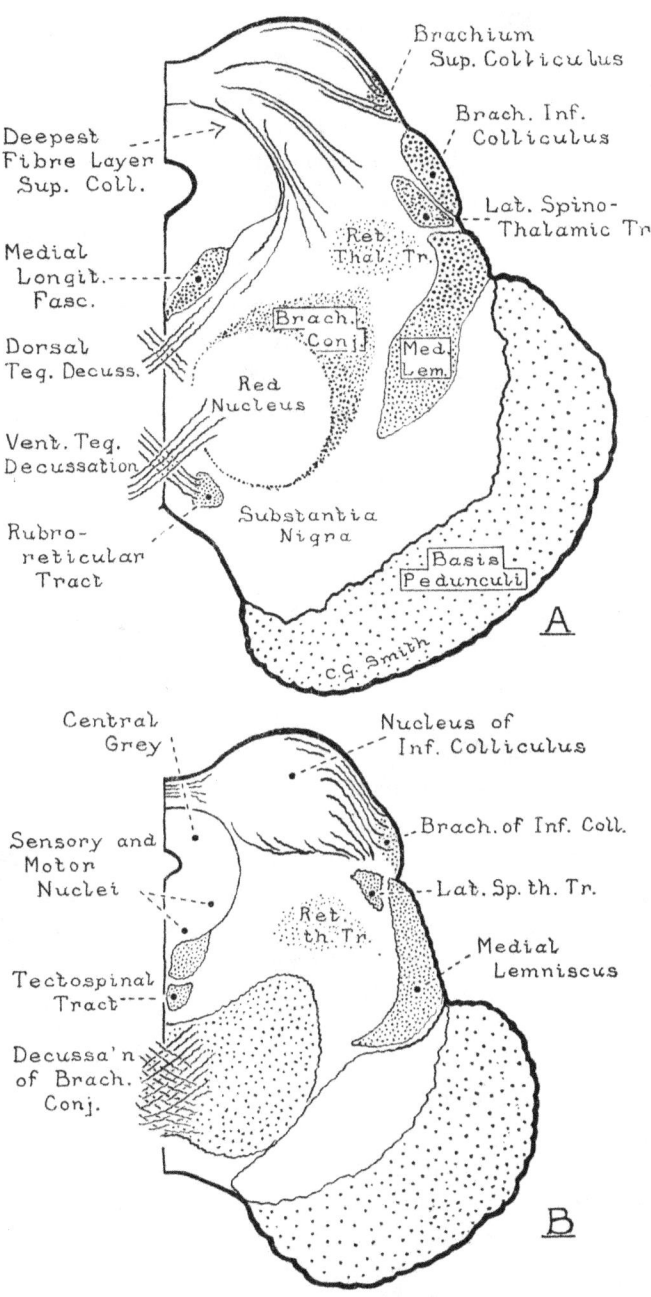

FIGURE 40. Cross-sections of the midbrain at the levels shown in figure 36C. A: Superior colliculus. B: Inferior colliculus.

2. THE NUCLEUS OF THE INFERIOR COLLICULUS

Dorsal to the central grey matter in the caudal half of the midbrain we find the nucleus of the inferior colliculus. It is an ovoid mass, encapsulated by the fibres of the lateral lemniscus which end in it. Its neurons serve as relay stations on diverging auditory pathways. Most of its fibres form the brachium of the inferior colliculus, which is the continuation of the auditory pathway to the forebrain. When this pathway is cut we lose our sense of hearing. Other fibres from other cells in this nucleus pass toward the tegmentum, and some pass upwards into the superior colliculus immediately above it. All these carry impulses that excite reflex automatic postural adjustments in response to auditory stimuli. A familiar example of this reflex postural adjustment is the movement of a cat's ears in response to sound. The pathways that descend to the motor nuclei are called the tectobulbar and tectospinal tracts. These come in part from the superior colliculus and will be described below.

3. THE LAMINATED GREY MATTER OF THE SUPERIOR COLLICULUS

The grey matter of the superior colliculus has three layers of cells, a superficial, a middle, and a deep layer. These layers of cells are imbedded in four layers of fibres, the deepest of which is one of efferent fibres, while the other three contain the afferent fibres of the superior colliculus. The afferents in layers 1 and 2—the superficial layers—are optic nerve fibres in brains of lower animals, but these are replaced in man by fibres from the cerebral hemisphere. The third layer contains afferent fibres that come to it from the medial lemniscus, from the lateral spinothalamic tract, and from the inferior colliculus. In this way the superior colliculus receives a wide variety of information and it uses this in its important role as a posture-controlling centre.

The efferent fibres of the superior colliculus form the deepest layer in the colliculus before they course ventrally around the central grey matter to reach the tegmentum. Some of them cross the midline to form a decussation, just ventral to the central grey matter, called the *dorsal tegmental decussation*. After they cross the midline they descend close to the midline in a bundle that distributes fibres to the motor nuclei of the brain and cord. They excite reflex responses to auditory and visual stimuli. These descending fibres take the names

MIDBRAIN

of their origin and termination, that is, they are called the tectobulbar (brain stem) and tectospinal tracts. Their course need not concern us.

4. THE RED NUCLEUS

The red nucleus is a conspicuous feature of the central part of the tegmentum of the midbrain. It is an ovoid mass of cells, circular in cross-section, that extends from the plane between the superior and inferior colliculi upwards to lie partly in the diencephalon. It gets its name from its pink colour in a freshly cut brain. It is pink because it contains more capillaries than its surroundings. The red nucleus is a relay station for impulses conveyed to it from many different parts of the nervous system but most of them arrive via the brachium conjunctivum. The cells give off fibres that form diverging pathways for the regulation of muscular activity. Some of these carry impulses up to the diencephalon to be relayed to the cerebral hemisphere. Others carry impulses caudally to reach the motor nerve nuclei of the brain and cord. Thus the red nucleus can influence motor activity in two ways: (1) directly, through its descending pathways to motor nuclei, and (2) indirectly, through pathways that ascend into the forebrain.

The fibres that descend from the red nucleus do not form compact bundles. They cross the midline at the level of the red nucleus in the ventral part of the tegmentum as the ventral tegmental decussation. They then descend close to the midline, a few of them going as far as the motor nuclei of the cord, but most of them ending in the reticular formation of the brain stem which will be described later.

5. THE SUBSTANTIA NIGRA

This is a thick layer of cells that extends the length of the midbrain. It may be likened to a pad placed between the basis pedunculi and the rest of the tegmentum. Its cells are remarkable for the black pigment they contain. This nucleus requires no stain: it stands out clearly in a cross-section of a freshly cut brain as a black band. It is present only in brains of animals having a basis pedunculi, suggesting that its cells are in large part a relay station for impulses arriving from the cerebral hemisphere. The substantia nigra also receives collaterals from the fibres in the adjacent medial lemniscus. The efferent fibres of the substantia nigra, that is, the fibres of the cell bodies in this nucleus, course in part into the adjacent tegmentum and in part they ascend to the cerebral hemisphere. The latter pathway suggests a feed-back mechanism, to influence activity in the cerebral hemisphere. The pathway to the tegmentum feeds impulses into the pathways that descend to motor nuclei from the reticular formation.

The function of the substantia nigra is the regulation of the flow of impulses into motor pathways. Degeneration of this nucleus is associated with an involuntary, maintained contraction of most of the muscles of the body. In such a case any movement is an effort. The condition is described as one of muscular rigidity.

6. THE FIBRE TRACTS OF THE MIDBRAIN

All but two of the fibre tracts of the midbrain that will be considered in this chapter were described with the external features. The two that remain are the lateral spinothalamic tract and the brachium conjunctivum.

a) THE LATERAL SPINOTHALAMIC TRACT

This tract courses along the dorsal border of the medial lemniscus medial to the brachium of the inferior colliculus. The latter serves as a landmark and has to be cut in making the incision to cut the pathway of pain at the level of the midbrain.

b) THE BRACHIUM CONJUNCTIVUM

This bundle courses ventrally and medially in its ascent from the cerebellum to the red nucleus. It is already ventral to the fourth ventricle before it leaves the pons segment and therefore it enters the tegmental part of the midbrain. The brachium conjunctivum then crosses the midline, the right and the left brachia passing through each other to form the large mass of decussating fibres at the level of the inferior colliculus. Two-thirds of the fibres end in the red nucleus, the rest sweep around and encapsulate it as they ascend to their relay station in the diencephalon.

6. The Diencephalon

I. DEFINITION

THE DIENCEPHALON of the mature brain is the portion of the forebrain that contains the third ventricle and lies between the right and the left cerebral hemispheres. Developmentally it includes a small part of the telencephalon (figure 6).

II. SHAPE AND SURFACES

The diencephalon has the shape of a wedge (figure 42). The edge of the wedge is its vertical anterior border. The base of the wedge faces backwards. Its superior and inferior surfaces are triangular and its right and left surfaces are quadrangular. The posterior surface becomes part of the superior surface because the border between them is rounded off. On this composite superior surface there is a broad, deep, median gutter that becomes progressively shallower and narrower toward its anterior end.

The inferior surface has three parts—an anterior, a posterior, and a middle part. The middle part is attached to the midbrain, the anterior and the posterior parts are free. The anterior part rests on the diaphragma sellae (figure 41). The posterior part is divided into right and left portions that overhang the superior colliculi (figure 42).

III. CAVITY

The cavity of the diencephalon is the third ventricle (figure 42). It is a median, cleft-like space with a roof, a floor, and an anterior wall. There is no posterior wall because the floor and roof converge toward the aqueduct which drains the third ventricle.

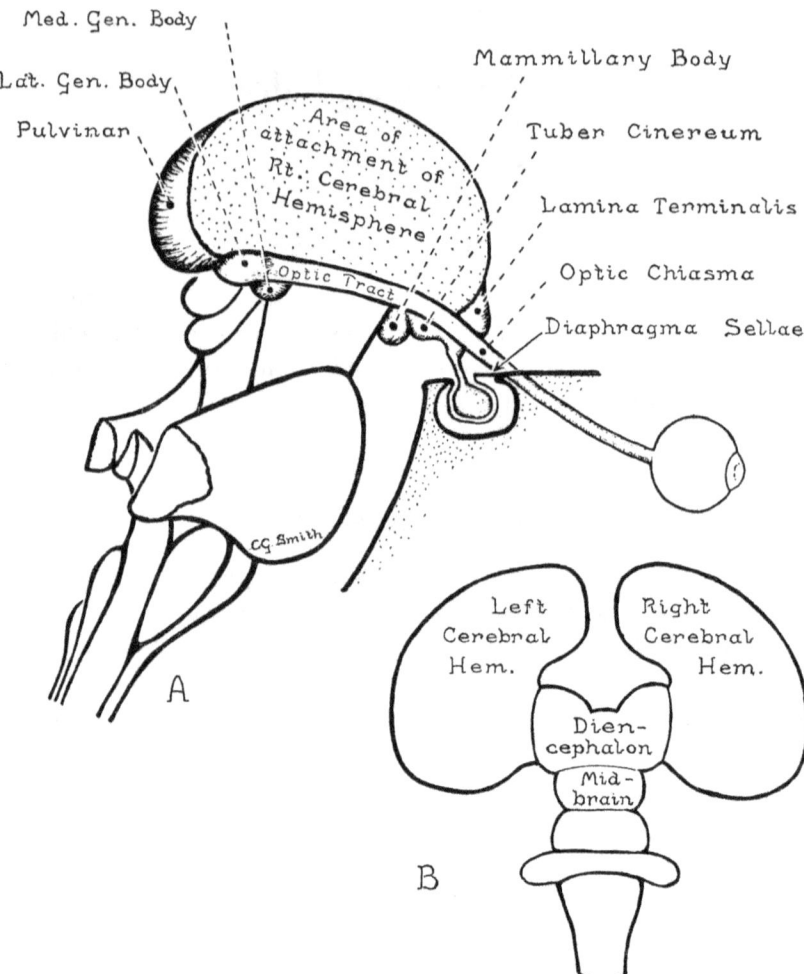

FIGURE 41. The relationships of the diencephalon. A: Lateral aspect of the brain stem. B: Dorsal aspect of the brain stem.

IV. THE SUBDIVISIONS OF THE DIENCEPHALON

The diencephalon as defined here is considered by some comparative neurologists to be the first part of the true brain to evolve. All its nuclei are association centres. There are no motor or sensory nuclei in the diencephalon as there are in the midbrain and the hindbrain. The latter segments of the brain may therefore be looked

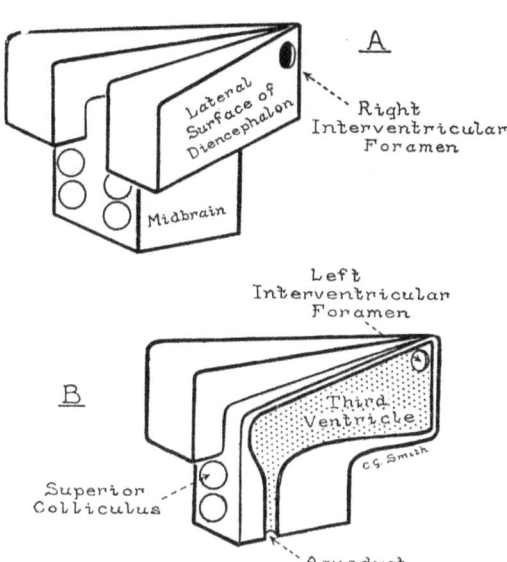

FIGURE 42. Schematic drawings of the diencephalon and midbrain. A: The dorsolateral aspect. B: A midsagittal section.

upon as modified portions of a primitive spinal cord. Since the diencephalon does not contain motor or sensory nuclei, no true nerves are attached to it. It is true that the optic nerves are attached to its ventral surface, but these only look like nerves—they are really attenuated parts of the brain.

Before the cerebral hemisphere develops, the nuclei of the diencephalon serve as the dominant association region of the central nervous system. When the cerebral hemisphere evolves and takes over this function, the nuclei of the diencephalon are woven into the pathways leading to and from the hemisphere. They retain their important association functions but now they are also, in a sense, all relay stations on pathways ascending to, or descending from, the cerebral hemisphere. The nuclei interpolated in ascending pathways (sensory) are in the dorsal part of the diencephalon and those in descending pathways (motor) are, with one exception, in the ventral part of the diencephalon. You will recall that the sensory and motor portions of the cord are similarly in a dorsal and a ventral location.

These relay nuclei of both the dorsal and the ventral parts of the diencephalon are in turn grouped according to their function to make up five subdivisions within the diencephalon. The largest of these is called the thalamus. The name means "a chamber" and is appropriate in the sense that it is a relay station, that is, an anteroom to the cerebral hemisphere. The other subdivisions of the diencephalon are named according to their position in relation to it. These subdivisions are represented in the diagrams of the diencephalon in figure 43.

A. THE SUBTHALAMUS

The subthalamus contains the relay nuclei of the motor pathways that are descending to striated muscles. These particular motor pathways are classified as extrapyramidal because they do not course through the pyramids as they pass through the medulla on their way to the spinal cord.

The subthalamus is a portion of the ventral part of the diencephalon. It is located directly above the tegmentum of the midbrain and contains the upper ends of the red nucleus and the substantia nigra, in addition to its own extrapyramidal relay nuclei.

B. THE HYPOTHALAMUS

The nuclei that make up the hypothalamus are relay stations on pathways descending to smooth muscle, to the heart, and to glands. This subdivision lies in front of the subthalamus in the ventral part of the diencephalon. Its inferior surface is free and rests on the diaphragma sellae.

C. THE THALAMUS

This is the largest subdivision of the diencephalon. It contains the relay nuclei of the general sensory pathway and also the relay nuclei of the feedback pathways that return information to the cerebral cortex after it has been processed in the cerebellum and in other parts of the brain, as will be described later.

It forms the major part of the dorsal (superior) portion of the diencephalon stretching its full length and resting in turn (from front to back) on the hypothalamus, the subthalamus, and the metathalamus.

D. THE METATHALAMUS

This subdivision contains two nuclei, one for the auditory pathway, and one for the optic pathway. The two nuclei are crowded onto the inferior surface of the posterior extremity of the thalamus, where they form an "after" portion of the thalamus, that is, a meta (after) thala-

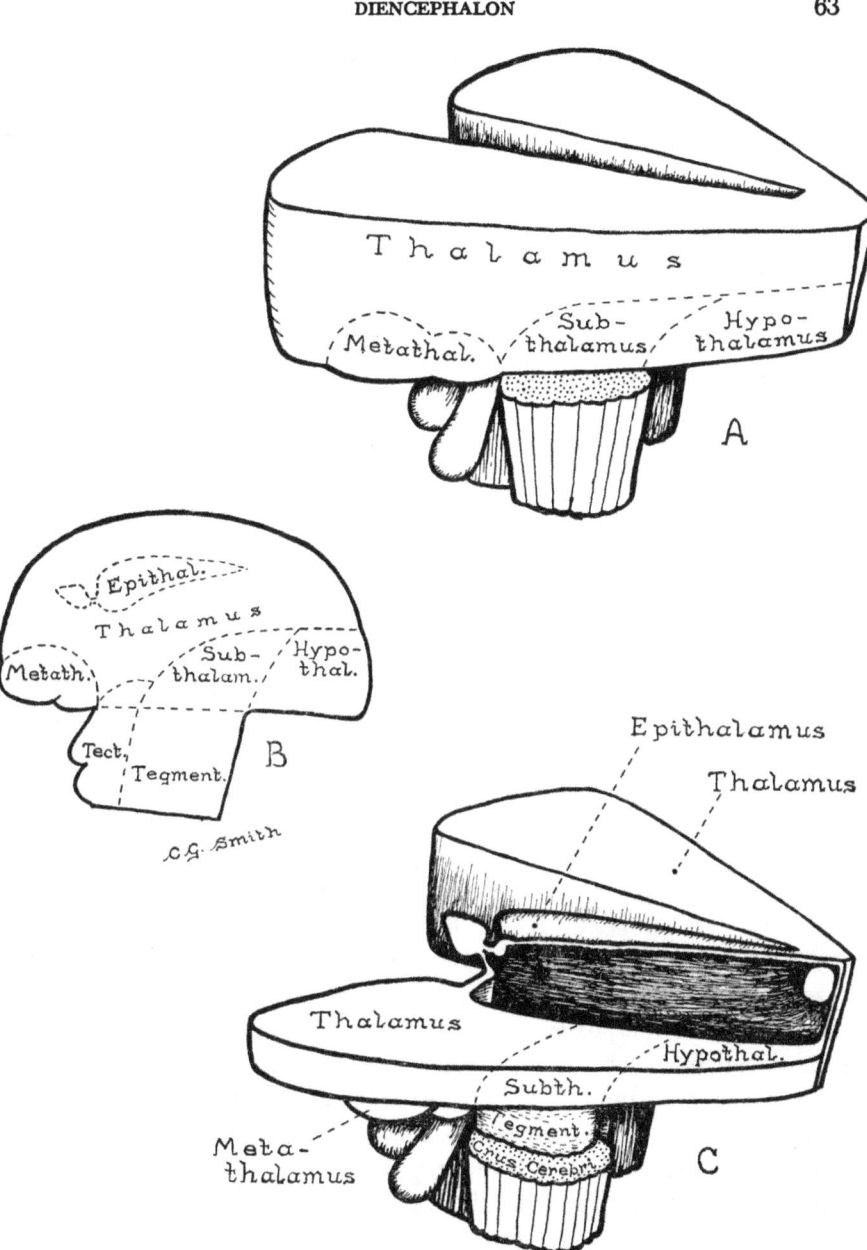

FIGURE 43. The subdivisions of the diencephalon and their relationships to the tectum, tegmentum and crus cerebri of the midbrain. A: Superior-lateral aspect of the diencephalon and midbrain. B: Subdivisions projected onto an outline of the lateral aspect of the diencephalon. C: The specimen shown in A after removing the upper part of the right half of the diencephalon and the upper part of the crus cerebri of the midbrain.

mus. They are of course a part of the posterior enlargement of the dorsal sensory portion of the diencephalon (see figure 43).

E. THE EPITHALAMUS

The ephithalamus is the smallest and the oldest portion of the diencephalon. Its nuclei form a slender elongated cell mass located alongside the roof of the third ventricle. The pineal gland, which is included in the epithalamus in this description, contains no nerve cell bodies and no nerve fibres other than those to its bloodvessels. It is a median conical structure attached to a band of transverse fibres that unite the right and the left epithalami at the posterior border of the membranous roof of the third ventricle.

The nuclei of the epithalamus are relay stations on pathways that come from the sense organs for smell and are descending to the tegmentum of the midbrain to initiate motor responses. As a relay station on a descending pathway, the epithalamus should be situated in the ventral part of the diencephalon with the subthalamus and hypothalamus.

It is a curious fact that in certain amphibians the right epithalamus is much larger than the left although the right and left olfactory sense organs are identical in size and structure.

V. THE EXTERNAL FEATURES OF THE DIENCEPHALON

A. THE SUPERIOR SURFACE

1. OUTLINE

The superior surface has the outline of a conventional valentine heart (figure 44). The apex is anterior and its notched base is posterior. At the floor of the posterior median notch the superior surface of the diencephalon is flush, and continuous, with the tectum of the midbrain. This junction is unmarked by an external feature and the transition zone is called the pretectal region. It is a significant region because it contains the relay nuclei of the pupillary constrictor pathway (constriction of the pupil in a bright light). To the right and left of the midline the diencephalon bulges backwards to overhang the corresponding superior colliculus. Along the right and the left borders are attached the cerebral hemispheres. The apex of the triangular superior surface lies between the two hemispheres but it is not free. It has the

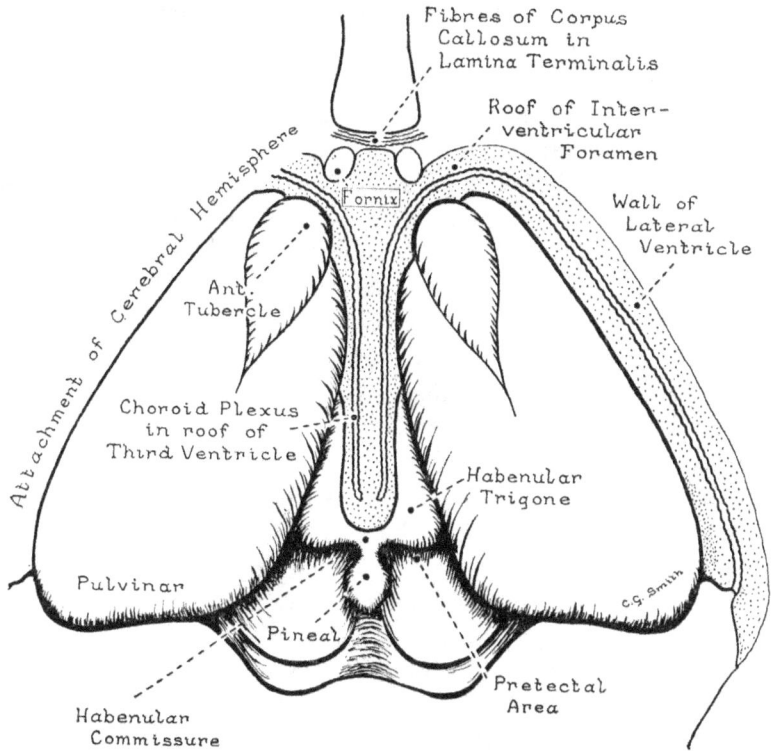

FIGURE 44. The features of the superior aspect of the diencephalon.

thin edge of a commissure attached to it. The latter, called the corpus callosum, connects the two hemispheres. Behind the thin edge of the corpus callosum, two bundles enter the superior surface of the diencephalon, one on each side of the midline. These come from the right and left hemispheres and each one is known as the fornix.

2. SURFACE FEATURES

The floor of the median gutter is formed by the membranous roof of the third ventricle and the epithalamus.

The membranous roof of the third ventricle is a narrow median band. At its anterior end it is attached to the back of the closely approximated right and left fornix bundles. Just behind this it has a lateral extension that is the roof of the interventricular foramen. Behind this it has the habenular trigone as its lateral border. This structure, and the habenular commissure, and the pineal gland to be described make up the epithalamus.

The habenular trigone is a slender ridge, so named from its fancied resemblance to the strap of a bridle (*habenula*, a strap). It is about 2 mm. wide at its posterior end and tapers to a point anteriorly a little short of the interventricular foramen. Its posterior end is connected with its fellow of the opposite side by a band of fibres called the habenular commissure. This commissure has the membranous roof of the third ventricle attached to its anterior border, and has the stalk of the pineal gland attached to its posterior border. The pineal gland is a cone-shaped mass about 8 mm. long that projects backwards above the superior colliculi in the midline.

The walls of the median gutter and the rest of the superior surface are formed by the right and the left thalamus. Each thalamus has a triangular outline. The rounded apex is called the anterior tubercle and has the membranous roof of the interventricular foramen attached to it. The base of the thalamus projects back to overhang the midbrain. This cushion-like mass is called the pulvinar (a cushion).

B. THE INFERIOR SURFACE
1. OUTLINE AND SURFACE FEATURES

The inferior like the superior surface has the outline of the conventional valentine heart (figure 45). The attachment of the midbrain takes up the middle and largest part of the inferior surface. In front of this is the free triangular ventral surface of the hypothalamus. The apex of this surface is formed by the anterior wall of the third ventricle called the lamina terminalis. Just behind this, the right and the left optic nerves are attached on each side. About half of the fibres of each nerve cross the midline here to form a transverse bar called the optic chiasma. These crossing fibres join uncrossed fibres and form a bundle that continues on as the optic tract. This courses back along the lateral border of the inferior surface of the diencephalon, hugging the lateral side of the crus cerebri (basis pedunculi). The optic tract fibres end in the lateral geniculate body, one of the nuclei of the metathalamus. This is a club-shaped swelling at the end of the optic tract located just lateral to the dorsolateral border of the crus cerebri. Its lateral part is superficial and bulges externally to form a conspicuous ovoid swelling. Its medial part bulges into the underside of the pulvinar and has only

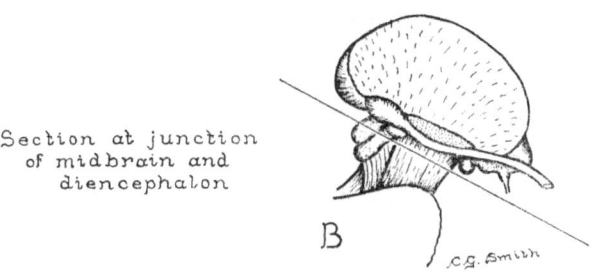
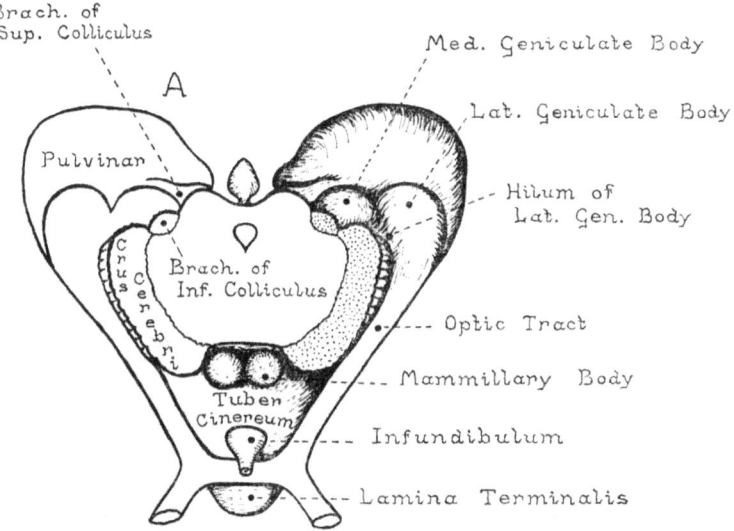

FIGURE 45. A: The features of the inferior aspect of the isolated diencephalon. B: The lateral aspect of the diencephalon and the midbrain.

a small external surface that forms the floor of a groove. This is the hilum of the lateral geniculate body.

On the medial side of the lateral geniculate body is the well defined swelling called the medial geniculate body (figure 45). This is the relay nucleus for the auditory pathway. It lies immediately dorsal to the crus cerebri, and at the end of the brachium of the inferior colliculus. The medial geniculate body is sometimes mistaken for the lateral geniculate body because the optic tract fibres that enter the hilum of the lateral geniculate body appear to continue into it. Behind the geniculate bodies is the underside of the pulvinar.

On the ventral surface of the hypothalamus immediately in front of the midbrain are two spherical masses, the right and left mammillary bodies. These are nuclei of the hypothalamus. Between the mammillary bodies and the optic chiasma the floor of the third ventricle is drawn down into a funnel-like projection which has the pituitary gland attached to its apex. The stem of this funnel-like projection is part of the neural lobe of the pituitary gland. It does not contain nerve cells. Its junction with the rest of the floor of the third ventricle, the part called the tuber cinereum (ashen-grey tubercle), is marked by a poorly defined encircling sulcus.

C. THE LATERAL SURFACE OF THE DIENCEPHALON

The lateral surface of the diencephalon serves as an area for the attachment of the cerebral hemisphere (figure 45B). Through this surface pass almost all the fibres that connect the brain stem and the cerebral hemisphere. They are called projection fibres and they may have their cell bodies in the hemisphere and convey impulses out of the hemisphere or have their cell bodies in the diencephalon and convey impulses into the hemisphere. These fibres, afferent and efferent, are crowded together in the stalk of the cerebral hemisphere just lateral to the diencephalon to form one compact fibre layer called the internal capsule. The reason for the crowding together of the fibres and the significance of the name internal capsule becomes clear if we trace, for example, the group of fibres that enter the hemisphere. The fibres leaving the hemisphere are forced to follow the same route but in the opposite direction.

The fibres that enter the stalk of the hemisphere through the lateral aspect of the diencephalon encounter and have to sweep around a large mass of cells in the hemisphere called the lentiform nucleus (figure 46). This nucleus as its name indicates has the shape of a biconvex lens. Its medial surface is opposed to the lateral surface of the diencephalon and about 1 cm. from it. Hence the fibres that enter the hemisphere from the diencephalon are deflected by it. They turn and course according to their destination in the hemisphere, forward, upward, backward or downward. In this way a lamina of closely crowded fibres is formed on the medial surface of the lentiform nucleus. It is called its internal, that is, its medial capsule.

Although the internal capsule is a portion of the cerebral hemisphere it is composed of pathways that pass through the diencephalon and it will be helpful therefore to describe it here.

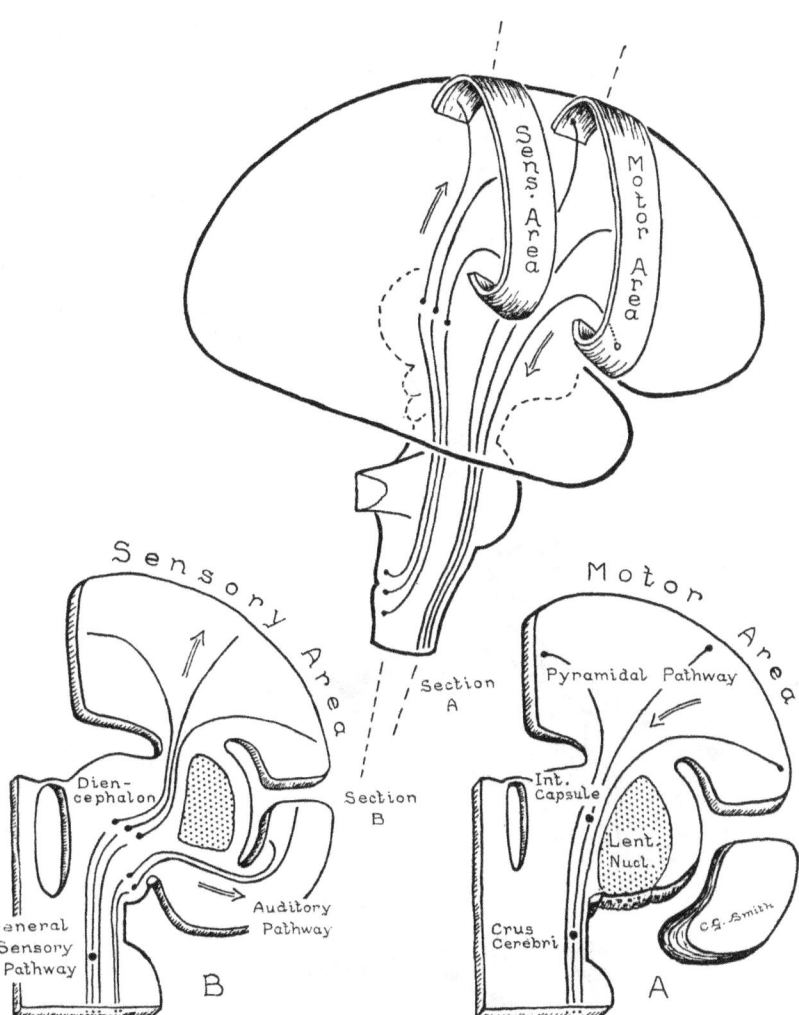

FIGURE 46. A diagram to show how the fibres of the sensory and the motor pathways are crowded together in the space between the diencephalon and the lentiform nucleus to form an internal (medial) capsule for the lentiform nucleus.

1. THE INTERNAL CAPSULE

a) DEFINITION

The internal capsule is a layer of projection fibres on the medial surface of the lentiform nucleus (figure 47). The internal capsule is only a part of the capsule of the lentiform nucleus and it is only internal in the sense that it is medial not lateral to it.

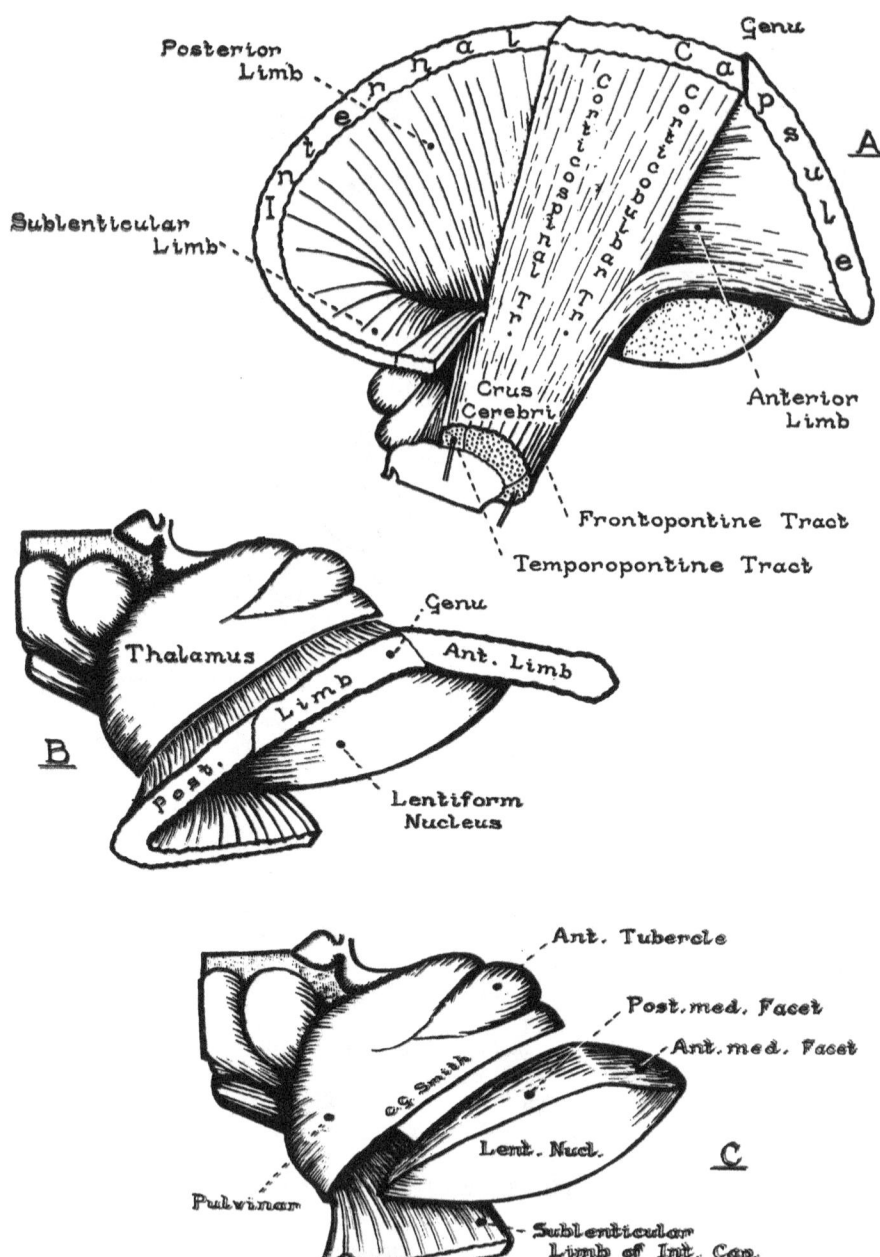

FIGURE 47. Drawings illustrating the form, subdivisions, and relationships of the internal capsule. A: Lateral aspect. B: Superior aspect plus diencephalon and lentiform nucleus. C: Same specimen as B after removing all the fibres of the internal capsule except those of the sublenticular limb.

b) Form and Subdivisions of the Internal Capsule

The form of the internal capsule may be likened to that of an open Japanese fan. Its handle is the crus cerebri (basis pendunculi). This consists of motor pathways that descend in the middle part of the internal capsule to reach the ventrolateral aspect of the midbrain (figure 47A). The concave aspect of the fan faces laterally and is applied to the lentiform nucleus.

The internal capsule has three triangular flattened portions that are the counterpart of three facets on the medial aspect of the lentiform nucleus. Of these three facets two are large and face anteromedially and posteromedially respectively. They meet to form the crest of an almost vertical ridge that extends upwards and forwards from a point near the middle of the inferior border of the nucleus (figure 47C). The flat portion of the internal capsule applied to the facet in front of this ridge is called the anterior limb of the internal capsule, the flat portion of the internal capsule behind the ridge is called the posterior limb, and the angle in the internal capsule opposite the ridge is called the genu. The third facet on the lentiform nucleus is on the posterior third of its inferior border. It faces inferiorly as well as medially. The part of the internal capsule applied to this inferior facet is appropriately called the sublenticular part or limb. It can be seen to advantage in figure 47C where all the parts of the internal capsule were removed except the sublenticular limb. The angle in the internal capsule between the sublenticular portion and the posterior limb is not named.

c) Relationships of the Internal Capsule to the Diencephalon

The lentiform nucleus is so placed within the stalk of the hemisphere and is so large that only its posteromedial facet is opposed to, that is, is lateral to the diencephalon (figure 47C). Hence only the posterior limb of the internal capsule and the genu are immediate lateral relationships of the diencephalon. The other two portions of the internal capsule of the lentiform nucleus are situated inside the cerebral hemisphere and intervene between the lentiform nucleus and some masses of grey matter that belong to the cerebral hemisphere. These will be described with the cerebral hemisphere.

d) The Location of the Motor Pathways within the Internal Capsule

The chief motor pathways in the internal capsule are the corticospinal, the corticobulbar, and the corticopontine tracts. The corticospinal tract is the pathway that initiates voluntary movements via the spinal nerves, the corticobulbar tract initiates voluntary movements via the cranial nerves, and the corticopontine tracts activate the cerebellum to co-ordinate the muscles involved in these movements.

The dissection illustrated in figure 47 shows the location of these tracts in the internal capsule and their course into the crus cerebri. The fibres that form the anteromedial border of the crus cerebri are coming from the anterior part of the hemisphere and help to form the anterior limb of the internal capsule. This bundle of fibres is the frontopontine tract. The fibres that form the posterolateral border of the crus cerebri are coming from the lateral part of the hemisphere and comprise the temporopontine tract. This is in the sublenticular limb of the internal capsule. The middle fibres of the crus cerebri are coming from the superior part of the middle portion of the hemisphere. They descend almost vertically and form the genu and the anterior half of the posterior limb. They are the fibres of the corticospinal and corticobulbar tracts plus a large number of corticopontine fibres. The latter come from the same part of the hemisphere as the pathways initiating voluntary movements. These descending pathways skim along the surface of the diencephalon, pressed against it by the lentiform nucleus (see figure 46) until they reach a point close to its lower border. Here they enter the diencephalon by passing deep to the optic tract (see figure 45). Their course through the diencephalon is as long as the optic tract is wide. At its lower border they emerge to form the crus cerebri of the midbrain.

D. THE MIDSAGITTAL SECTION OF THE DIENCEPHALON. THE THIRD VENTRICLE

The diencephalon like the other segments of the brain stem consists of right and left identical portions. In the diencephalon, however, these two portions are almost completely separated by the cavity of the diencephalon—the third ventricle.

The form of the third ventricle is shown in the schematic midsagittal section of the diencephalon in figure 42. Its detailed form and the features of its wall are shown in figure 48. In addition to its right and left walls it has a superior wall or roof, an inferior wall or floor, and an anterior wall. It has a short posterior wall, dorsal to the aqueduct, but this is considered to be a part of the roof.

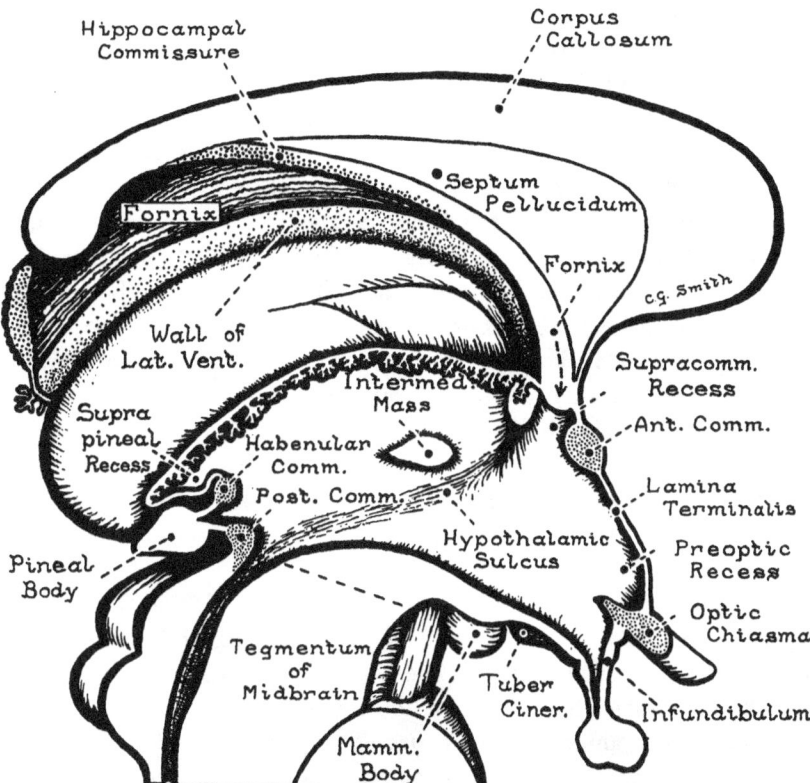

FIGURE 48. A midsagittal section of the midbrain, diencephalon, and the commissural band connecting the cerebral hemispheres to show the form and relationships of the third ventricle.

1. THE FLOOR OF THE THIRD VENTRICLE

The posterior half of the floor rests on the tegmentum of the midbrain because the diencephalon sits on the midbrain like the crosspiece of the letter "T." As already pointed out, the structures found in the tegmentum of the midbrain continue vertically upwards into the diencephalon to form the subthalamus. As they ascend, however, they move away from the midline and the central cavity of the midbrain expands ventrally between them to form the posterior part of the third ventricle.

Just a short distance in front of the attachment of the midbrain to the diencephalon, the floor of the third ventricle is formed by the medial parts of the mammillary bodies. In front of these again, the floor is pulled down to form a funnel-shaped evagination. Only the stem of this funnel is called the infundibulum (funnel). It is a part of the neural lobe of the hypophysis and it contains no nerve cell bodies. The rest of the funnel-like evagination is a part of the hypothalamus called the tuber cinereum (ash-grey tubercle) and it does of course contain nerve cells. An indistinct, shallow sulcus encircles the junction of the infundibulum and the tuber cinereum. In front of the tuber cinereum the floor of the ventricle is pulled upwards and backwards by the optic chiasma and its right and left connections with the optic tracts. The midsagittal section of the optic chiasma is shaped and hangs like a pear, its long axis vertical with a slight inclination forward and downward. This is an anatomical fact to be remembered because the crossing fibres at its lower edge are coming from the inferior nasal quadrants of the retina. They would be the first ones to be interrupted, therefore, by a tumour expanding upwards and backwards from the sella turcica.

2. THE ANTERIOR WALL OF THE THIRD VENTRICLE

The anterior wall of the third ventricle is formed by the lamina terminalis, the rostral border of the embryonic neural tube. It is a thin membrane which stretches from the lower part of the front of the chiasma to the anterior commissure. The pocket-like space between the chiasma and the lamina terminalis is called the preoptic recess. The anterior commissure is a bundle of fibres, about 2 mm. in diameter, which connects the parts of the right and the left cerebral hemispheres that lie deep to the temporal bones. It arches medially from the temporal lobe to reach the lamina terminalis and uses it as a bridge to cross the midline. It is of interest that the larger part of the anterior commissure in man is a new structure that connects parts of the brain that are concerned with memory. In lower animals its fibres connect the olfactory bulbs, the recipients of the olfactory nerves.

Above the anterior commissure there is a small supracommissural recess and above this, at the junction of the anterior wall and the roof of the third ventricle, the wall is formed by the fibres of the thin edge of the corpus callosum. Immediately behind this, the fornix of the right hemisphere and the fornix of the left hemisphere are pressed against each other to form a small part of the wall. The corpus callosum consists of commissural fibres that connect the two cerebral hemispheres. The fornix is a bundle of fibres that courses from the posterior part of the hemisphere to enter the anterior end of the diencephalon on its way to the mammillary body.

3. THE ROOF OF THE THIRD VENTRICLE

The larger part of the roof is membranous. It is a paper-thick membrane formed by the ependymal epithelium of the wall of the third ventricle and a supporting layer of neuroglia. It extends from the fornix in front, to the habenular commissure. It is attached on each side to the medial margin of the habenular trigone and in front of this structure it is directly continuous with the roof of the interventricular foramen. The roof of the interventricular foramen is in turn continuous with the membranous part of the medial wall of the lateral ventricle. The roof of the third ventricle is not stretched taut between its lateral attachments. It is ballooned dorsally to fill partially the deep median gutter on the dorsal surface of the diencephalon. This dorsal expansion of the third ventricle projects back over the pineal gland to form the suprapineal recess, recognizable in the X-ray examination of the air-filled ventricle. In spite of its dorsal evagination, the whole length of the roof of the ventricle has a linear choroid plexus on either side of the midline. Anteriorly each plexus is prolonged laterally into the roof of the interventricular foramen to be continuous with the choroid plexus in the wall of the lateral ventricle.

The posterior, smaller part of the roof of the third ventricle is formed by the walls of the hollow stalk of the pineal gland and the posterior commissure. The posterior commissure is at the entrance to the cerebral aqueduct. The pointed extension of the third ventricle contained in the stalk of the pineal gland is the pineal recess. The posterior commissure contains the crossing fibres of the light reflex pathway. Both pupils contract when a pencil of light strikes the temporal half of the retina of one eye. The habenular commissure usually has wart-like masses on its ventricular surface. These are never found on the posterior commissure. They are usually calcified and show up in X-ray examinations. Their significance is not known.

4. THE LATERAL WALL OF THE THIRD VENTRICLE

The right and left walls of the third ventricle may be 1 to 10 mm. apart. Each wall has three features, the interventricular foramen, the hypothalamic sulcus, and the intermediate mass. The latter is a mass of cells uniting the right and the left thalami.

a) THE INTERVENTRICULAR FORAMEN

This is a short canal, 3 to 5 mm. in diameter. Its medial end is located in the angle between the roof and the anterior wall of the third ventricle. The fornix forms its anterior and its inferior borders, the anterior tubercle of the thalamus forms its posterior border. Its roof is membranous and the choroid plexus it contains almost fills the opening.

b) THE HYPOTHALAMIC SULCUS

This is a poorly defined groove which extends from the floor of the interventricular foramen to the aqueduct. It divides the lateral wall of the ventricle into dorsal and ventral halves. The dorsal part is the thalamus proper, the ventral part consists of (*a*) the hypothalamus anteriorly, and (*b*) the subthalamus posteriorly. A line parallel to the lamina terminalis and passing through the posterior border of the mammillary body corresponds roughly to the junction of the two ventral portions of the diencephalon.

c) THE INTERMEDIATE MASS

This is an adhesion between the right and left walls of the third ventricle. It unites the right and left thalami proper, and is located just above the hypothalamic sulcus, anterior to its midpoint. It may be thread-like or it may have a diameter equal to the ventricular surface of the thalamus proper, that is, a diameter of about 15 mm. This adhesion is apparently without functional significance. It is only present in about two-thirds of the brains examined at autopsy.

VI. THE INTERNAL STRUCTURE OF THE DIENCEPHALON

We will begin the study of the internal structure of the diencephalon using the dissection illustrated in figure 49. This preparation is a dissection of the medial aspect of the specimen illustrated in figure 48, that is, the medial aspect of the left half of the forebrain. To prepare it a cut is first made across the junction of the midbrain and the diencephalon as deep as the fibres of the crus cerebri and then the diencephalon is peeled away from the posterior limb of the internal capsule. In doing this the optic tract and its nucleus of termination, the lateral geniculate body, break away and are left in place because the optic tract passes lateral to the crus cerebri. To complete the

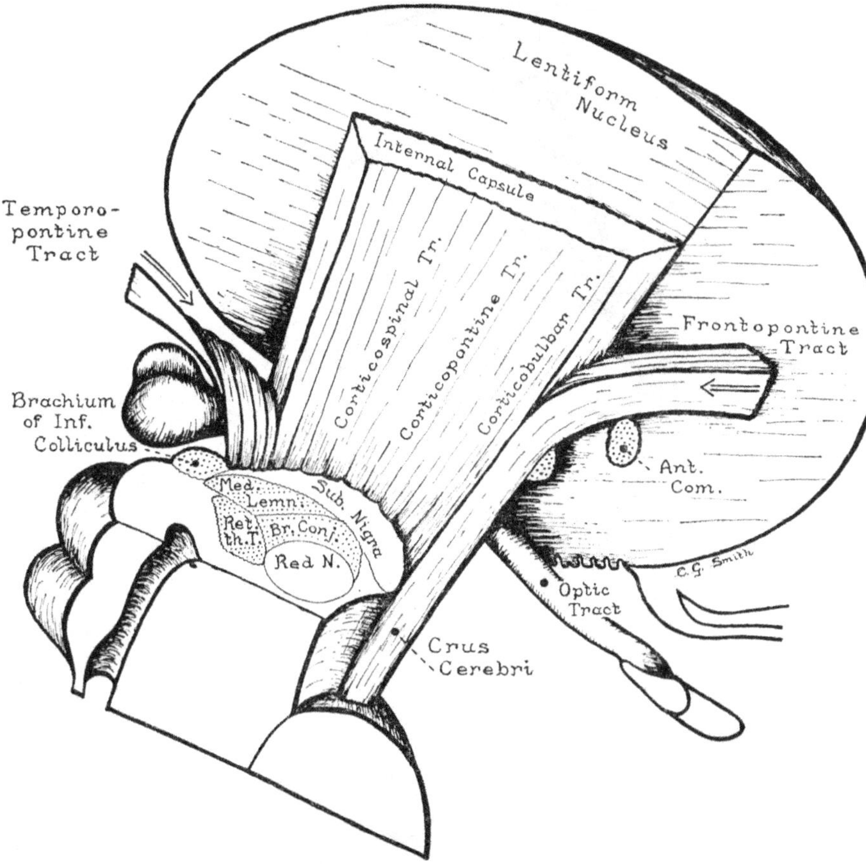

FIGURE 49. A preparation showing the upper aspect of the left half of the tectum and tegmentum of the midbrain, the medial aspect of those fibres of the internal capsule that descend into the crus, the optic tract and its nucleus of termination and the medial aspect of the lentiform nucleus.

dissection, all parts of the cerebral hemisphere are removed except the lentiform nucleus and those portions of the internal capsule that contain fibres of the crus cerebri.

Let us reconstruct the left half of the diencephalon subdivision by subdivision using this preparation as a foundation.

A. THE SUBTHALAMUS
1. FORM AND RELATIONSHIPS

The subthalamus is, in part, an extension of the tegmentum of the midbrain. It rises to the level of the hypothalamic sulcus, and extends

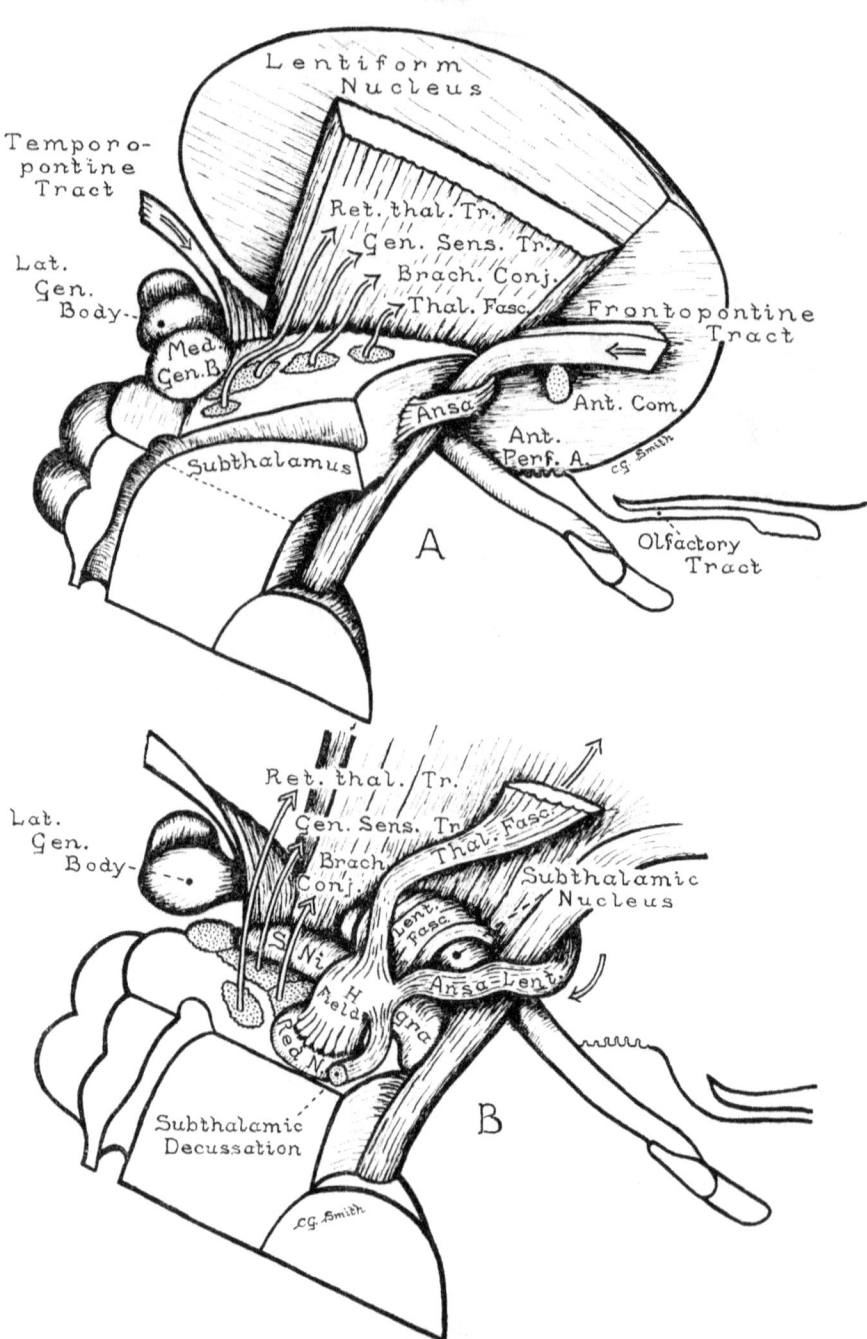

FIGURE 50. A: The specimen illustrated in figure 49 after adding the subthalamus, the ansa lenticularis and the medial geniculate body. B: The above specimen dissected to show the chief nuclei and fibre tracts of the subthalamus.

forwards in the wall of the third ventricle as far as a line drawn parallel to the lamina terminalis and just caudal to the mammillary body (figure 52A). Laterally the subthalamus is applied to the posterior limb of the internal capsule (figure 50A) and reaches as far forward as its anterior border. The anterior surface of the subthalamus faces anteromedially and is applied to the hypothalamus. The latter tapers to a thin edge posteriorly in the wall of the third ventricle.

2. INTERNAL STRUCTURE OF THE SUBTHALAMUS

The posterior part of the subthalamus contains three bundles of fibres that are ascending to reach the overlying thalamus (figure 50). The posterolateral bundle is the general sensory pathway which includes the medial lemniscus and the spinothalamic tracts. The posteromedial bundle is the reticulothalamic tract. The anterior bundle (labelled "Brach. Conj." in figure 50A) is the dento-rubro-thalamic pathway, which includes fibres of the brachium conjunctivum and fibres from cells in the red nucleus.

The anterior, larger part of the subthalamus contains the relay stations and fibre bundles of the pathways descending from the lentiform nucleus. Most of these pathways continue on into the tegmentum of the midbrain via the bundle labelled "H field" in figure 50B. A smaller number pass superiorly into the anterior part of the thalamus via the thalamic fasciculus; and a much smaller number (as illustrated in the frontal section of figure 51) pass medially into the hypothalamus. Each of the three efferent fibre bundles of the subthalamus carries impulses that excite or modify movements involving striated muscles. Let us locate first the nuclei of the subthalamus and then trace the fibre bundles. In tracing these we will describe them as they might be revealed in a dissection, keeping in mind that the pathways they contain may be interrupted by synapses in one or several of the nuclei that they encounter in the subthalamus.

3. NUCLEI OF THE SUBTHALAMUS

The upper end of the ovoid red nucleus and the upper border of the plate-like substantia nigra project into the subthalamus from the midbrain (figures 49 and 50B). Both have been described with the tegmentum of the midbrain as relay stations on extrapyramidal pathways. The substantia nigra extends up almost to the level of the lower border of the optic tract. Here it is replaced by the subthalamic

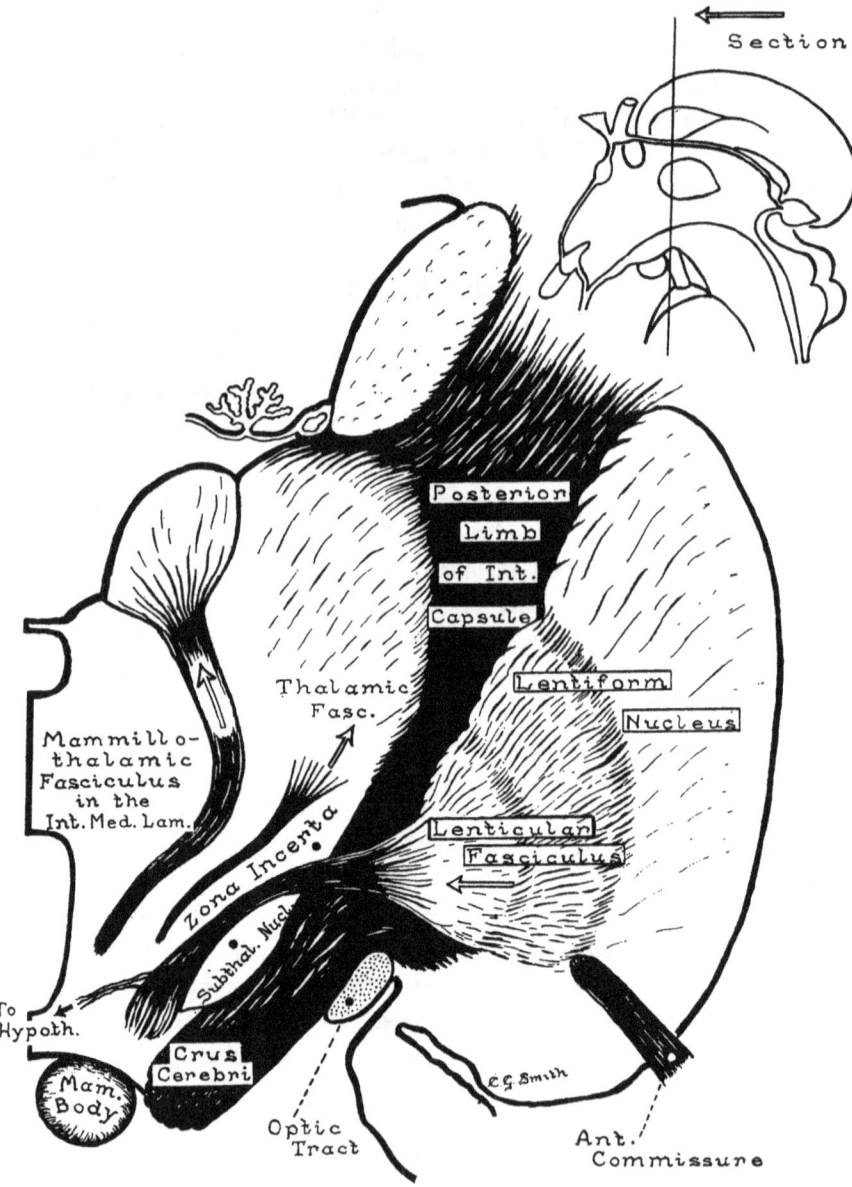

FIGURE 51. A frontal section of the diencephalon, internal capsule and lentiform nucleus in the plane shown in the upper right diagram.

nucleus, a flattened ovoid mass of unpigmented cells that lies along its upper border. The fourth nucleus of the subthalamus, the zona incerta, cannot be demonstrated by dissection because it is so small and its cells do not form a compact mass. It is shown in the frontal section illustrated in figure 51 as a plate of cells between the capsule of the subthalamic nucleus and the layer of fibres labelled "Thalamic Fasciculus."

4. FIBRE BUNDLES OF THE SUBTHALAMUS

Two fibre bundles enter the subthalamus from the lentiform nucleus. One of these—the lenticular fasciculus—takes a direct route by piercing the posterior limb of the internal capsule to enter the lateral surface of the subthalamus. The other bundle—the ansa (loop) lenticularis—makes a detour around the anterior border of the internal capsule to enter the subthalamus through its anterior surface (figure 50A). The fibres of both these fasciculi pass across the medial aspect of the subthalamic nucleus and form a part of the fibre capsule of this nucleus.

At the upper pole of the red nucleus a large number of these fibres cross the midline. This subthalamic decussation is one of the sites where pathways descending to striated muscles from the lenticular nucleus can cross the midline. Hence disease of one lentiform nucleus or one of the nuclei of the subthalamus of one side is followed by disturbances of function on the opposite side of the body. The symptoms and signs may take the form of "rigidity"—a maintained contraction of agonists and antagonists; or a tremor—a rhythmic alternating contraction of agonists and antagonists; or spontaneously occurring, unwilled movements.

The densely packed interweaving fibres at the rostral pole of the red nucleus form a crest or cap for the red nucleus, a Haubenfeld (cap plus field). This is usually designated **Field H of Forel**. It is in a sense the part of subthalamus in which pathways group themselves into bundles according to their destination. One of these is **the thalamic fasciculus**. It is destined for the anterior end of the thalamus and courses forward and superiorly as a flattened bundle separated from the fibre capsule of the subthalamic nucleus by the grey matter called the zona incerta (figure 51). Another fasciculus—**the hypothalamic fasciculus**—contains only a few fibres. It passes medially into the portion of the hypothalamus that projects back on the medial side of the subthalamus. The largest efferent bundle of the subthalamus descends into the tegmentum of the midbrain passing through and around the red nucleus. These fibres convey their impulses to the nuclei of the reticular formation and through them to the motor nuclei.

B. THE HYPOTHALAMUS

1. FORM AND RELATIONSHIPS
(Figures 48, 52, 53)

The hypothalamus is wedge-shaped (figure 52). Its anterior vertical border, the lamina terminalis, is the thin edge of the wedge. The inferior surface of the hypothalamus is triangular and free. The apical part of this surface is formed by the lamina terminalis plus a small portion of the grey matter of the anterior perforated area on either side of it. Behind this, from front to back, we find in order the optic chiasma, the tuber cinereum with the stalk of the hypophysis attached to it, and the mammillary bodies. The lateral border of the inferior surface is the optic tract. The medial surface forms that part of the wall of the third ventricle that lies ventral to the hypothalamic sulcus, and in front of a vertical line behind the mammillary body. The lateral surface has the cerebral hemisphere attached to it. It extends backwards from the lamina terminalis to the anterior border of the posterior limb of the internal capsule and is directly continuous with the grey matter of the olfactory portion of the lentiform nucleus. Through this surface the hypothalamus receives many fibres from the cerebral hemisphere. These are irregularly spaced and help to obscure the poorly defined line of junction of hypothalamus and hemisphere.

The superior surface is triangular. Its apical part is crossed by the anterior commissure that connects the right and the left hemispheres. Behind the commissure the fornix enters the superior surface of the hypothalamus and immediately behind this again is the floor of the interventricular foramen. The rest of the superior surface of the hypothalamus carries the anterior portion of the thalamus.

2. THE NUCLEI OF THE HYPOTHALAMUS

The hypothalamus is divided into regions and these in turn into smaller portions that are called nuclei or areas, depending on whether the cells are in a cluster or dispersed (figure 53). A description of the location of these regions, nuclei, and areas will be useful in interpreting reports of experimental work.

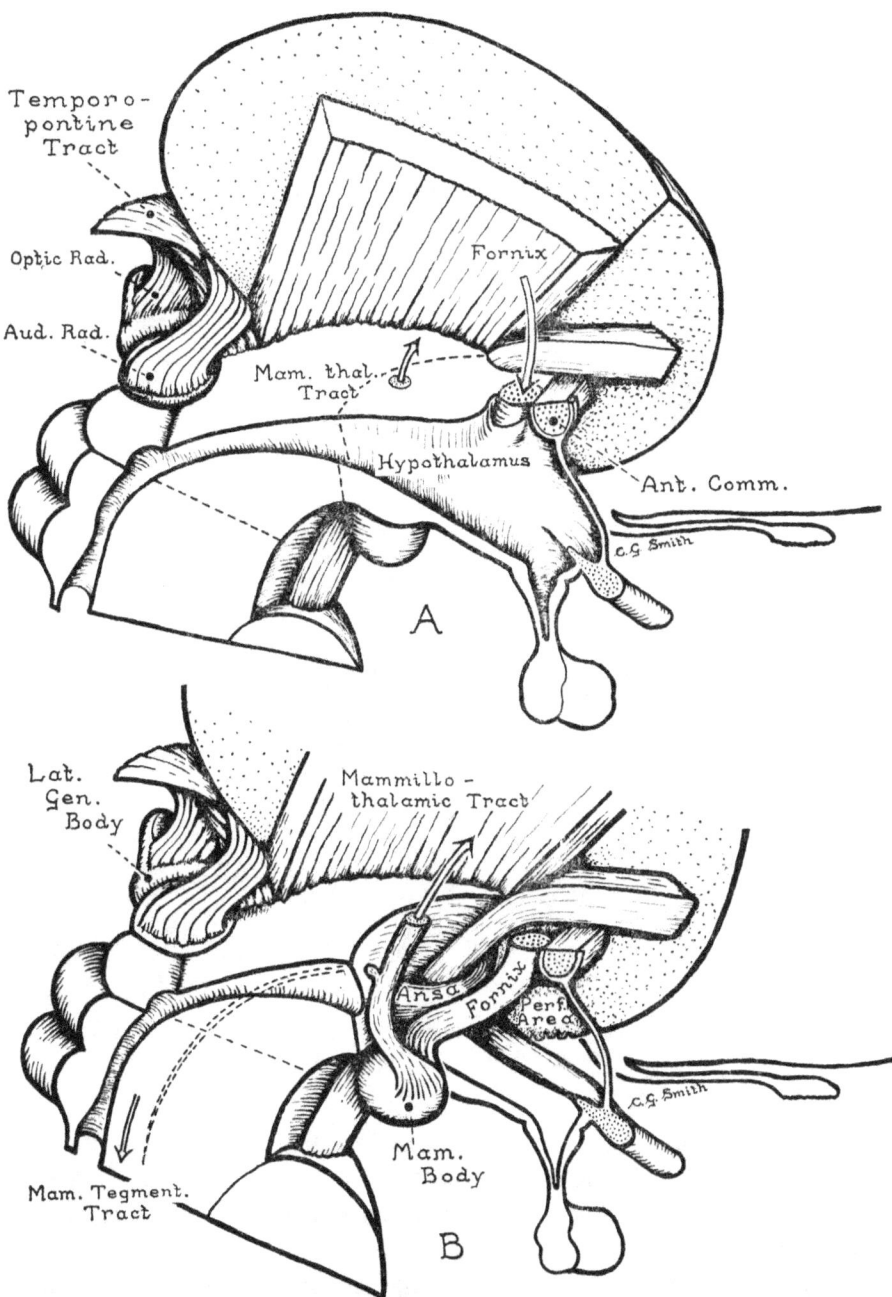

FIGURE 52. A: The specimen illustrated in figure 50A after adding the hypothalamus and the optic and auditory fibres that help to form the sublenticular limb of the internal capsule. B: The above specimen dissected to show the large fibre tracts of the hypothalamus.

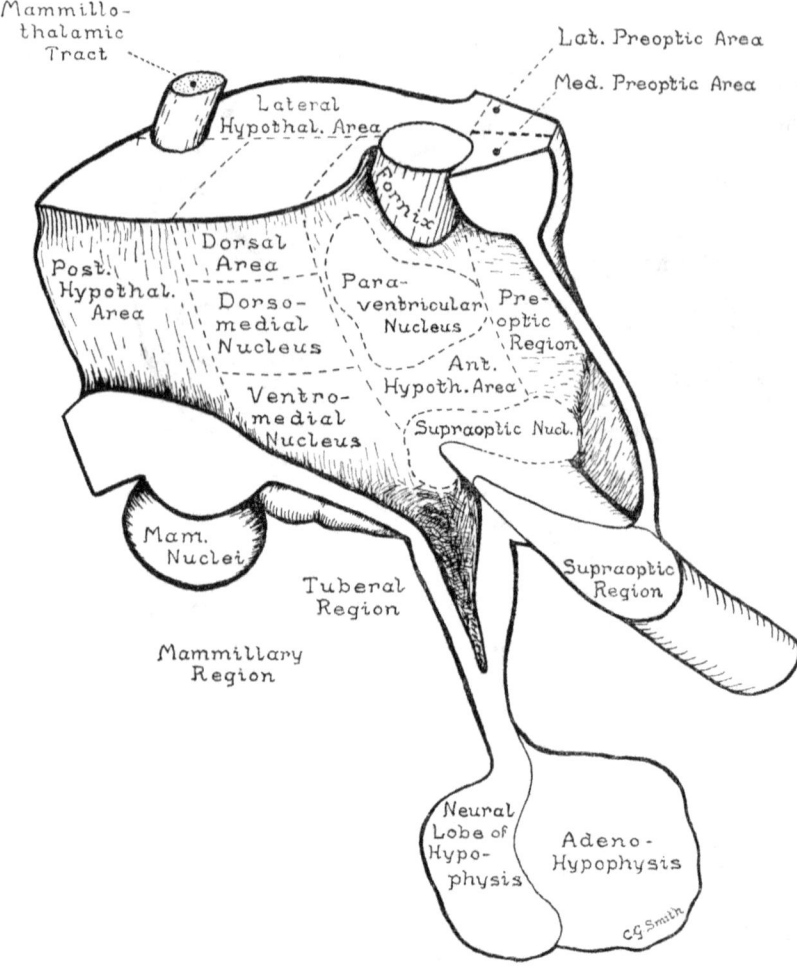

FIGURE 53. The ventricular and superior aspects of the isolated hypothalamus and the hypophysis of the left side.

The hypothalamus has a lateral portion containing many longitudinal fibres and a medial portion that is relatively free of these fibres. The parasagittal plane between the medial and lateral portions contains the terminal part of the fornix and the mammillo-thalamic tract. The medial portion is divided into four regions from front to back. Each of these is given the name of the external feature at its ventral border. Above the optic chiasma and the beginning of the optic tract is the **supraoptic region**. In front of this is a small part called the **preoptic region**. Above the tuber cinereum is the **tuberal region** and above the mammillary body is the **mammillary region**.

The parts of the supraoptic region are (*a*) a paraventricular nucleus, applied to the medial side of the fornix, (*b*) a supraoptic nucleus applied to the medial side of the lower end of the optic tract, and (*c*) a mass of scattered cells comprising the anterior hypothalamic area.

The tuberal region also has three parts, a ventral, a dorsal, and an intermediate. The ventral and intermediate portions are the ventromedial and dorsomedial nuclei of the hypothalamus. The scattered cells in the dorsal part make up the dorsal area.

The mammillary region has a ventral and a dorsal portion. The ventral part contains the mammillary nuclei. The cells of the dorsal part make up the posterior hypothalamic area.

The lateral part of the hypothalamus has a small anterior and a large posterior portion. The anterior small portion comprises the lateral preoptic region. The posterior portion is the lateral hypothalamic area. It contains many large cells whose fibres descend into the brain stem.

3. FIBRE BUNDLES OF THE HYPOTHALAMUS DEMONSTRABLE BY DISSECTION

It is remarkable that a subdivision of the diencephalon with a maximum dimension of 15 mm. should have within it fibre bundles that are large enough to dissect. Of these the largest is the fornix. It enters the superior surface of the hypothalamus between the anterior commisure and the interventricular foramen and takes a direct route to the lateral side of the mammillary body. The upper part of its course is marked by a ridge on the wall of the third ventricle. The fibres of the fornix synapse with cells in the medial part of the mammillary body. Here a compact bundle of fibres called the mammillo-thalamic tract begins. It passes upward arching first backwards and then forwards to reach the anterior part of the thalamus. Just as it bends forward it gives off a small bundle called the **mammillo-tegmental tract** that descends through the subthalamus to reach the tegmentum of the midbrain. This efferent fibre bundle of the hypothalamus contributes to the cap-like fibre mass of the red nucleus as it passes through the subthalamus.

4. THE FUNCTION AND MECHANISM OF THE HYPOTHALAMUS

The hypothalamus integrates complex patterns of response that are mediated chiefly by visceral effectors, that is, through smooth muscle, heart, and glands. Examples of such complex responses are—emotional responses and those involved in maintaining the body temperature, the proper water balance, and blood pressure. Striated muscles are also involved in some of these responses, for example, smiling as part of an emotional response, shivering as part of the response to raise the body temperature, and accelerated rate of respiration as part of the response to lower the body temperature. Hence the hypothalamus must have descending pathways to the preganglionic nuclei and to the motor nuclei of striated muscles as well.

Just as the hypothalamus must have motor pathways to the effectors it must be informed of changes in the environment and changes in mental activity. This information may be brought to it by nerve fibres or by physical and chemical changes in the blood passing through it.

Since the hypothalamus effects its responses chiefly through the autonomic nervous system which is divided into sympathetic and parasympathetic parts it might be anticipated that the hypothalamus is a dual mechanism. The evidence suggests that in a general way, the anterior portion initiates parasympathetic activity and the posterior part initiates sympathetic activity. Experimental stimulation of the anterior part excites responses such as slowing of the heart, constriction of the pupils, and peristalsis, whereas stimulation of the posterior part excites the opposite responses. The difference in function between the anterior and posterior portions of the hypothalamus holds true for temperature regulation. It is suggested that certain nerve cells in the anterior portion are stimulated by a rise in temperature of the blood and in turn initiate responses such as vasodilation, increased respiration, and sweating. On the other hand certain nerve cells in the posterior portion are stimulated by a lowering of the temperature of the blood and excite responses such as vasoconstriction and a rapid pulse.

5. THE EFFERENT PATHWAYS OF THE HYPOTHALAMUS

The efferent pathways of the hypothalamus are of two kinds. One is a neural pathway conducting nerve impulses. The other is a pathway conducting hormones that are elaborated within the cytoplasm of the nerve cells of the hypothalamus.

a) THE EFFERENT PATHWAYS DISCHARGING NERVE IMPULSES

(1) **To Preganglionic Nuclei**

The following three pathways are available:

(*a*) **Via periventricular fibres and the dorsal** (note, not the medial) **longitudinal fasciculus.** This pathway leaves the medial part of the hypothalamus. Unmyelinated fibres course medially to the wall of the third ventricle and then turn caudally descending in the floor of the aqueduct and then in the floor of the fourth ventricle. These fibres carry impulses to the preganglionic nuclei of nerves 3, 7, 9, and 10.

(*b*) **Via a pathway that leaves the lateral part of the hypothalamus.** Fibres descend from the lateral hypothalamic area and pass through the subthalamus into the lateral part of the reticular formation of the brain stem. They may descend all the way to the preganglionic nuclei of the cord or be interrupted by a synapse in the reticular formation. This is the course of the pupil dilator pathway which is interrupted in Horner's Syndrome and is characterized by a unilateral small pupil, drooping eyelid, and hot, dry skin of the head and neck.

(*c*) **Via the mammillo-tegmental tract.** This tract leaves the mammillo-thalamic tract and descends through the subthalamus into the reticular formation of the midbrain. Reticular spinal fibres may complete the pathway to the preganglionic nuclei of the brain stem and the spinal cord.

(2) **To the Thalamus**

The following two pathways are available: (*a*) via periventricular fibres and (*b*) via the mammillo-thalamic tract.

(*a*) **Via periventricular fibres.** These fibres course medially and then superiorly into the medial part of the thalamus. From here the impulses may be relayed to the anterior part of the cerebral hemisphere.

(*b*) **Via the mammillo-thalamic tract.** Fibres of the mammillo-thalamic tract ascend to the anterior part of the thalamus, which relays them to the medial part of the cerebral hemisphere.

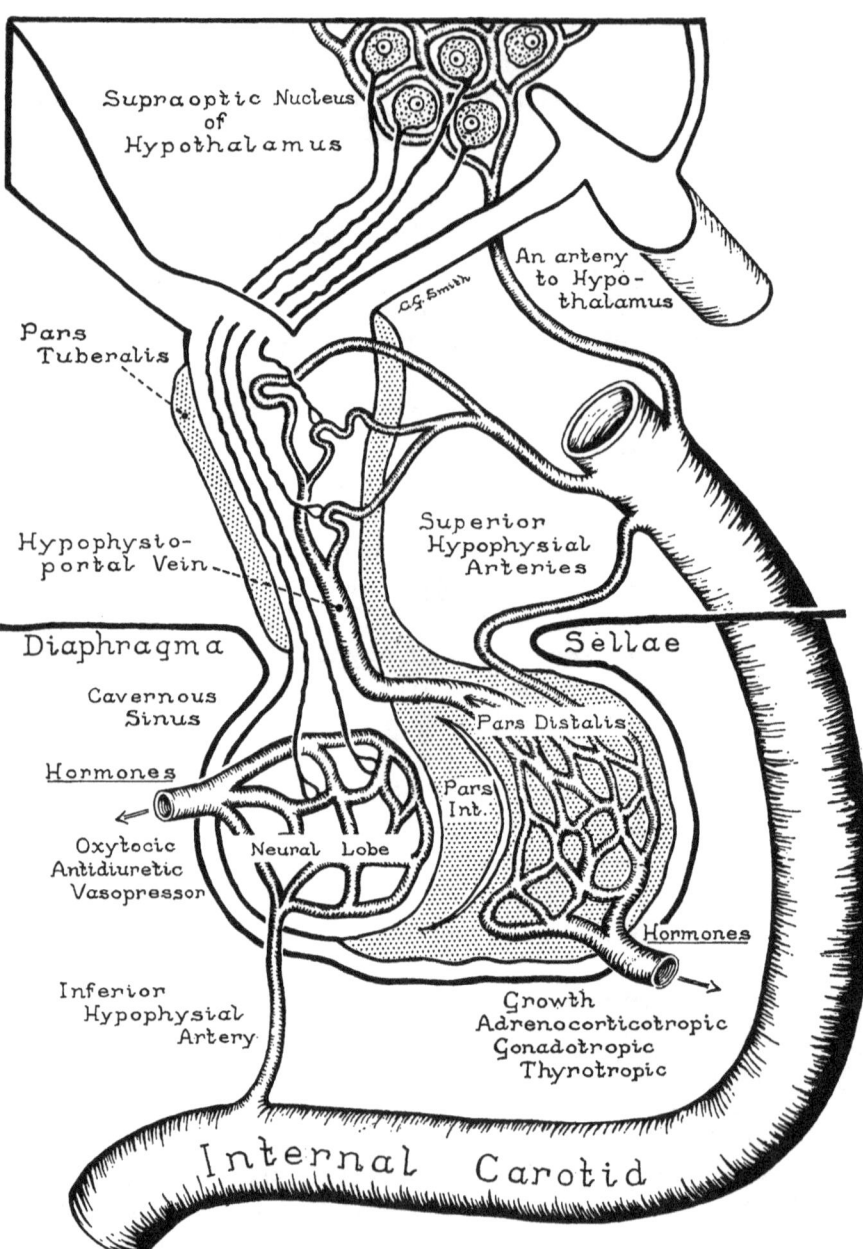

FIGURE 54. A diagrammatic sagittal section of the hypophysis and the floor of the third ventricle showing the subdivisions of the hypophysis, the supraopticohypophysial tract and their blood supply.

These two pathways from the hypothalamus to the thalamus and thence to the cerebral hemisphere may excite the perception of emotional feelings.

b) THE EFFERENT PATHWAYS DISCHARGING HORMONES

Figure 54 is a schematic drawing of a midsagittal section of the floor of the third ventricle and the hypophysis to show how hormones that are manufactured within the cells of the hypothalamic nuclei can find their way into the capillaries of the hypophysis.

The portion of the hypophysis that is derived from the floor of the third ventricle is called the neural hypophysis. It has a slender, elongated, funnel-like stem and a bulb-like extremity. A shallow encircling sulcus marks the junction of the stem and the tuber cinereum. The glandular part of the hypophysis, the adenohypophysis, the part derived from Rathke's Pouch, forms (a) the pars distalis, (b) the pars intermedia, and (c) the pars tuberalis.

Hormones discharged into the capillaries of the neural hypophysis are manufactured in the cytoplasm of the nerve cells of the supraoptic nucleus, the paraventricular nucleus, and in some cells of the tuber cinereum. These hormones flow into the axons which serve as a tubular passage conducting them either into the stem of the hypophysis where some end, or beyond this into the bulb-like terminal part where most of them end. At both sites the nerve fibres discharge their hormones into the lumen of a capillary but the capillaries of the stem of the hypophysis and of the bulb-like extremity are independent of one another. Those of the bulb-like extremity drain directly into the cavernous sinus, but those of the stem drain into a vein which, like the portal vein of the liver, carries the blood into another set of capillaries. The second set of capillaries is in the adenohypophysis. Here, the hormones that entered the blood in the neural hypophysis may activate the cells of the adenohypophysis. In this way the hypothalamus may regulate the release of gonadotrophic and thyrotrophic hormones from the anterior lobe.

One of the hormones discharged into the capillaries of the bulb-like extremity of the neurohypophysis is the antidiuretic hormone. This is manufactured in the cells of the supraoptic nucleus in response to changes in the osmotic pressure of the blood that supplies them. Hence, when the blood requires to be diluted, a supply of antidiuretic hormone is formed and released to promote the reabsorption of the water lost from the blood in the glomeruli of the kidney.

6. THE AFFERENTS OF THE HYPOTHALAMUS

The stimuli activating the hypothalamus are partly provided by the changes in the blood that is flowing through it, and partly by nerve impulses. Since the hypothalamus is known to control all visceral responses the sources of its afferent fibres may in part be predicted. Obviously it should receive data from the sense organs of the body. This data may arrive via the reticulothalamic tract which it is known can be activated by any sense organ of the body except the olfactory. The data from the olfactory sense organ arrives via the anterior perforated substance of the cerebral hemisphere. Secondly, the hypothalamus should receive processed data from the association regions of the thalamus and of the cerebral hemisphere. This information may be conveyed from the thalamus via fibres that descend close to the wall of the third ventricle. The information from the hemisphere may reach it via the anterior perforated substance or via the fornix or the stria terminalis. The last two bundles enter the hypothalamus from above at its anterior end. They will be described when the cerebral hemisphere is studied.

C. THE METATHALAMUS

The lateral and medial geniculate bodies make up the metathalamus. These two nuclei are located side by side on the inferior aspect of the pulvinar and serve as relay stations along the special sense pathways from the eye and the ear, respectively. The lateral geniculate body receives the optic tract at its anterior end; the medial geniculate body receives the brachium of the inferior colliculus, that is, the auditory pathway at its postero-medial border.

The cells of the lateral geniculate body are arranged in six layers parallel to the inferior surface. The optic fibres enter the clefts between the cell layers anteriorly. The efferent fibres course laterally between the cell layers to emerge from the lateral side of this nucleus and continue on into the sublenticular part of the internal capsule. This part of the optic pathway is called the optic radiation.

The cells of the medial geniculate body form a well circumscribed cluster. The fibres leaving the nucleus and going to the cerebral hemi-

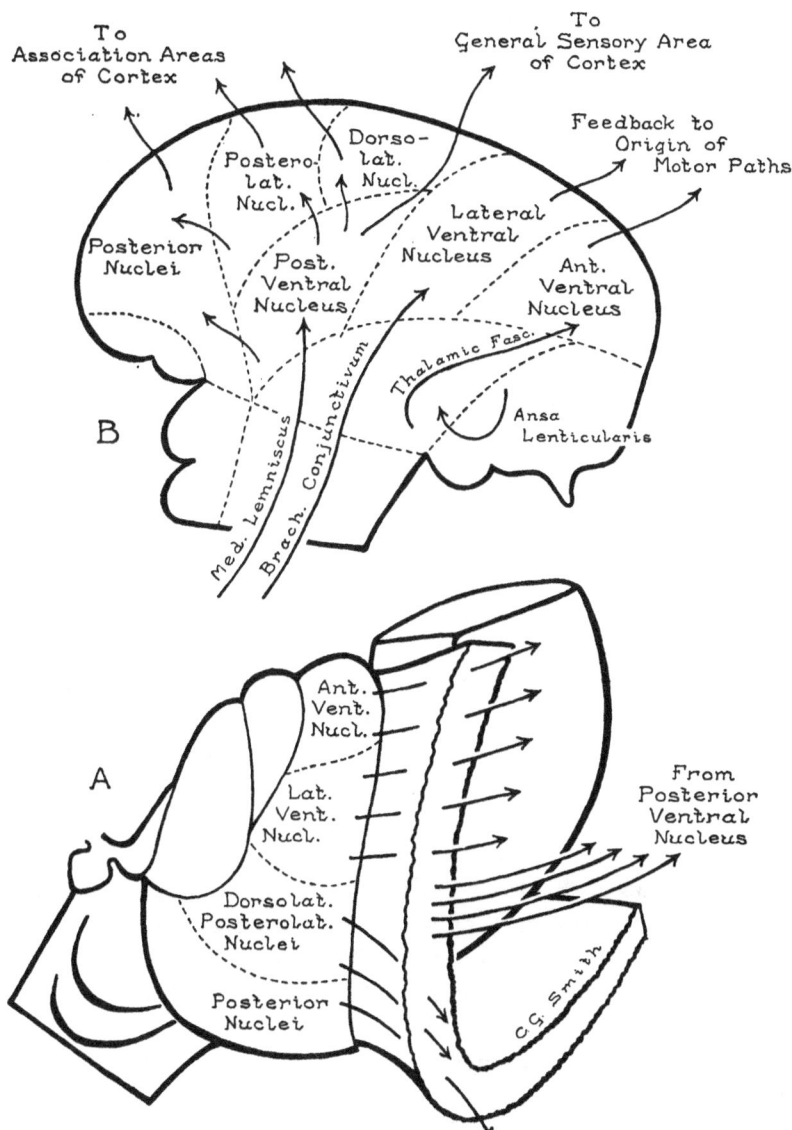

FIGURE 55. A: The superior aspect of the diencephalon (right half) to show how the posterior limb of the internal capsule is formed, in part, by fibres from each of the lateral nuclei of the thalamus. B: A schematic sagittal section of the diencephalon and midbrain to show the connections of the lateral nuclei of the thalamus.

sphere emerge from its superior aspect and course laterally above the optic pathway to help form the sublenticular limb of the internal capsule. This bundle is called the auditory radiation.

The relationships of the optic radiation and the auditory radiation are shown in figures 52 and 57. In figure 57 the sublenticular limb and the posterior limb of the internal capsule are shown with their full complement of fibres. Included in these is the brachium of the superior colliculus. The fibres of this bundle are descending from the visual area of the cerebral hemisphere to reach the superior colliculus to excite automatic movements of the eyes. They mingle with the fibres of the optic radiation in the sublenticular part of the internal capsule and then course medially within the diencephalon above both the optic and the auditory pathways to emerge from the inferior surface of the pulvinar just medial to the medial geniculate body.

D. THE THALAMUS

1. FORM AND RELATIONSHIPS

This very large subdivision is an ovoid mass that forms the superior portion of the diencephalon and extends its full length. Inferior to it from front to back are in turn the hypothalamus, the subthalamus, and the metathalamus (figure 43). Medial to it is the third ventricle and the habenular trigone. Lateral to it is the genu and the posterior limb of the internal capsule (figure 47).

2. INTERNAL STRUCTURE

The thalamus is divided into three unequal portions, an anterior, a medial and a lateral, each of which contributes to the formation of the superior surface (figure 56). The anterior portion forms the anterior tubercle of the thalamus. It is a conical or horn-shaped mass that lies along the upper border of a vertical partition of fibres that divides the thalamus into medial and lateral portions. This partition—called the internal medullary lamina—is prolonged backward beyond the pointed posterior end of the anterior tubercle for a short distance and then becoming less distinct curves toward the midline in the transverse vertical plane of the habenular commissure. Thus the lateral portion of the thalamus expands medially at the posterior end of the diencephalon to form the whole of the pulvinar.

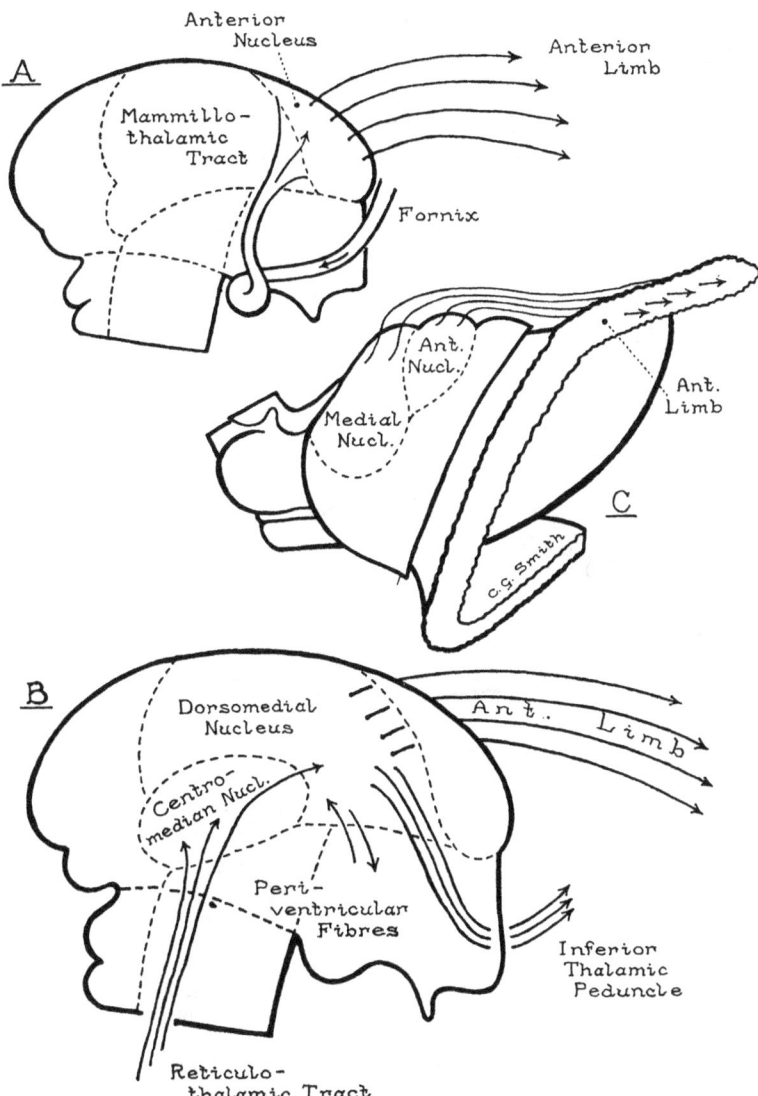

FIGURE 56. A: A schematic sagittal section of the diencephalon and midbrain to show the connections of the anterior nucleus of the thalamus. B: A schematic sagittal section of the diencephalon and midbrain to show the connections of the medial nuclei of the thalamus. C: The superior aspect of the diencephalon (right half) to show how the anterior limb of the internal capsule is formed, in part, by fibres of the anterior and medial thalamic nuclei.

94   BASIC NEUROANATOMY

a) THE LATERAL THALAMIC NUCLEI

The nuclei of the lateral portion of the thalamus and their chief afferents can be charted in a schematic sagittal section of the lateral portion of the diencephalon and the adjoining midbrain (see figure 55B). In such a section the subthalamus and the hypothalamus lie

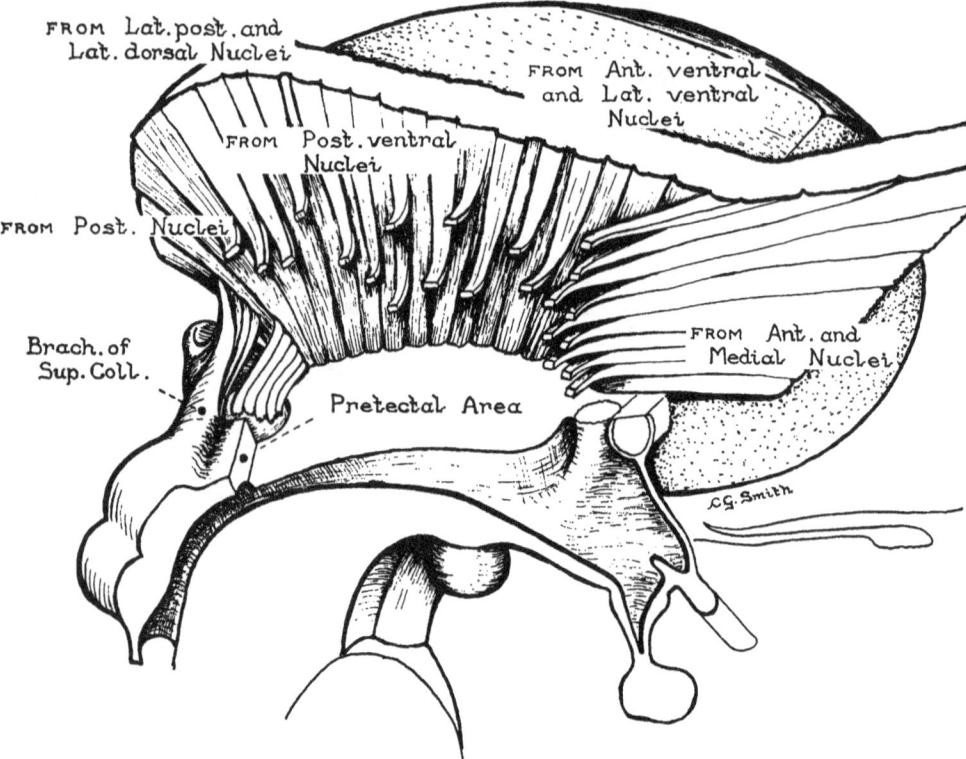

FIGURE 57. The medial aspect of the internal capsule showing the fibres contributed by each of the thalamic nuclei.

below a line that arches from the upper end of the plane separating tectum and tegmentum in the midbrain to reach the middle of the anterior border of the diencephalon. The thalamus forms all the rest of the section except for a small portion on the underside of the pulvinar which represents the metathalamus. Within the thalamus the outlines of its five groups of nuclei are shown. Let us consider first the location and the afferents of each group of nuclei and then describe their efferent fibres.

(1) **Location and Afferents of Each of the Groups of Nuclei in the Lateral Part of the Thalamus**

(*a*) **The posterior nuclei.** These are the nuclei of the pulvinar. They are located behind the transverse vertical plane of the habenular commissure. In a sagittal section through the lateral part of the brain stem this plane coincides approximately (see figure 55B) with the plane between the tectum and the tegmentum of the midbrain. The posterior nuclei of the lateral thalamus are connected with the tectum of the midbrain by a slender isthmus formed by the pretectal region. The posterior nuclei, as their restricted relationships would suggest, receive most of their afferents from the thalamic nuclei located in front of them. These afferents convey partially processed data to be relayed to the association areas of the cortex.

(*b*) **The posterior ventral nuclei.** These are directly above the posterior part of the subthalamus through which ascends the fibre bundle containing the medial lemniscus and the spinothalamic tract. This nucleus receives the bundle through its inferior surface and relays the impulses it carries—from the skin and muscles—to the cerebral cortex. The arrival of these impulses in the cerebral cortex excites the appropriate sensation, that is, either touch, pain, warmth, cold or a sense of position.

(*c*) **The lateral ventral nucleus.** The name of this nucleus does not fit its position as seen in a sagittal section. It is located in front of the posterior ventral nuclei and above the part of the subthalamus that gives passage to the brachium conjunctivum. It receives this pathway from the cerebellum through its inferior surface and serves as its relay station to the cerebral cortex. The impulses it sends to the cortex reach the area of cortex that gives origin to the motor pathways. Hence, along this pathway the cerebellum can influence the successive stages of a voluntary movement by relaying impulses to the area of cortex initiating voluntary movement.

(*d*) **The anterior ventral nucleus.** This nucleus is located above the anterior end of the subthalamus where the thalamic fasciculus ascends into the thalamus. The thalamic fasciculus conveys impulses from the lentiform nucleus. The anterior ventral nucleus relays these impulses to the area of cerebral cortex that gives origin to the motor pathways. Hence the lentiform nucleus can, like the cerebellum, influence the successive stages of the performance of a voluntary movement by

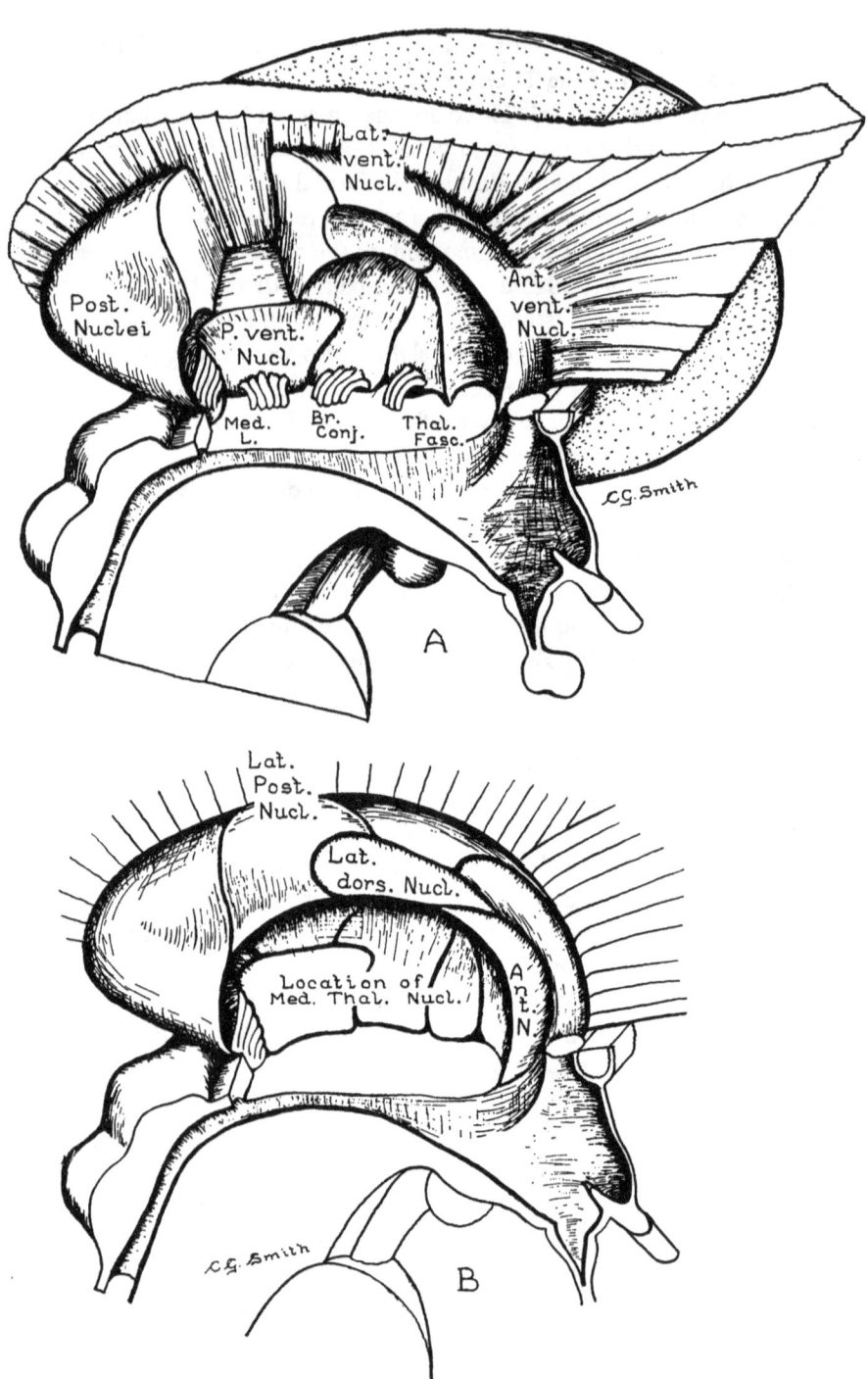

FIGURE 58. The form and relationships of the thalamic nuclei. A: The three ventral lateral thalamic nuclei and the posterior thalamic nuclei in position on the medial surface of the internal capsule. B: The same preparation with the addition of the lateral posterior nucleus, the lateral dorsal nucleus and the anterior thalamic nucleus.

relaying impulses to the area of cortex initiating that voluntary movement.

(e) **The dorsolateral and posterolateral nuclei.** These are functionally similar and are treated as a unit here. The form and location of each nucleus within this group is shown in the dissection of the diencephalon illustrated in figure 58 (B). The composite mass is located directly above the posteroventral nuclei and just in front of the pulvinar. These nuclei receive all their afferents from the other nuclei of the thalamus and relay the processed data to association areas of the cerebral cortex. These nuclei, phylogenetically, antedate those of the pulvinar, which develops as a subdivision of this nuclear group.

(2) **The Efferent Fibres of the Lateral Nuclei of the Thalamus**

The efferent fibres of the lateral nuclei of the thalamus, as already indicated, enter the cerebral hemisphere. The efferent fibre bundle from each nucleus helps to form the internal capsule as illustrated in figures 55 and 57. Note how each nucleus in turn, from the anterior end of the thalamus to the posterior end contributes the fibres of the successive parts of the posterior limb and the posterior part of the sublenticular limb of the internal capsule.

At the anterior end of the thalamus, the anterior ventral nucleus and the lateral ventral nucleus send their fibres laterally to form the genu and the anterior half of the posterior limb. These fibres to the cerebral cortex are mingling in the internal capsule with the descending pathways that initiate voluntary movement. This is fitting because they are ascending to the area of cortex in which the motor pathways begin.

The part of the posterior limb directly behind the motor pathways receives the fibres from the posteroventral nuclei. This is the relay station for the general sensory pathway. Behind these in turn are the fibres from the dorsolateral and posterolateral and the posterior nuclei of the pulvinar. These groups of fibres pass laterally and form the rest of the posterior limb and the adjacent part of the sublenticular limb.

b) THE MEDIAL THALAMIC NUCLEI

Just as a schematic sagittal section through the lateral part of the diencephalon was used to chart the lateral nuclei and the afferents reaching them through the lateral part of the subthalamus, so a similar sagittal section through the medial part of the diencephalon may be used to chart the medial and anterior nuclei and the afferents reaching them from the medial parts of the subthalamus and the hypothalamus. Such a section showing the connections of the medial thalamic nuclei is illustrated in figure 56B. It shows that the medial nuclei are located directly above the subthalamus and the posterior part of the hypothalamus. Behind the medial nuclei are the nuclei of the pulvinar, and in front of them are the anterior thalamic nuclei.

The medial group of thalamic nuclei includes several small nuclei and two large ones. Only the two large nuclei will be described. They are named from their positions the dorsomedial and the centromedian nuclei.

(1) **Location of the Medial Nuclei of the Thalamus and Their Afferents**

(*a*) **The centromedian nucleus.** The centromedian nucleus is an ovoid mass having its long axis in the sagittal plane and resting on the posterior medial part of the subthalamus. Here the reticulothalamic tract enters it and the centromedian nucleus is in part a relay station for this pathway.

(*b*) **The dorsomedial nucleus.** The dorsomedial nucleus forms the very large dorsomedial portion of the medial nuclear mass. It is resting on the anterior part of the subthalamus, through which it receives some reticulothalamic fibres, and on the hypothalamus through which it receives some periventricular fibres. It is in contact with each of the lateral thalamic nuclei as well as the anterior nuclei and it is from these that the dorsomedial nucleus receives most of its afferent fibres.

(2) **The Efferent Fibres of the Medial Nuclei of the Thalamus**

The efferent fibres of the medial nuclei carry impulses to the cerebral cortex, to the lentiform nucleus, and to the hypothalamus.

(*a*) **Fibres to the cerebral cortex.** Most of the fibres leaving the medial thalamic nuclei and going to the cerebral cortex come from the dorsomedial nucleus. A large number of these pass forward to reach the cortex of the anterior part of the hemisphere and help to form the anterior limb of the internal capsule. These are the fibres that a lobotomy operation is designed to cut. A smaller number of fibres may go to the cerebral cortex on the inferolateral part of the surface of the hemisphere (insular area). These fibres help to form a bundle called the inferior thalamic peduncle, which loops anteriorly, inferiorly and laterally around the anterior border of the internal capsule to enter the hemisphere through the anterior perforated substance. As it enters the hemisphere it is just ventral to the ansa lenticularis.

(*b*) **Fibres to the lentiform nucleus.** The fibres that leave the medial nuclei and end in the lentiform nucleus form the inferior thalamic peduncle.

(*c*) **Fibres to the hypothalamus.** The fibres to the hypothalamus descend close to the wall of the third ventricle as part of the thin periventricular lamina of fibres.

In summary, the connections of the medial nuclei suggest that the medial thalamic nuclei help to process somatic and visceral data and pass this on to (*a*) the cerebral cortex for further processing, (*b*) the lentiform nucleus to modulate motor activity of striated muscles, and (*c*) the hypothalamus to modulate visceral activity.

c) THE ANTERIOR THALAMIC NUCLEI

The horn-shaped mass of cells containing the anterior thalamic nuclei forms the anterior-superior border of the dorsomedial nucleus (figure 56A). Its afferent fibre bundle is the mammillothalamic tract which ascends from the hypothalamus within the internal medullary lamina.

The efferent fibres of the anterior nuclei pass forward to reach the cortex of the anteromedial part of the hemisphere and they help to form the anterior limb of the internal capsule (figures 56 and 57). They may convey information that is interpreted by the cortex as emotional feeling.

E. THE EPITHALAMUS

The nuclei of the epithalamus are in the habenular trigone. The afferent fibres of these nuclei form the slender fasciculus called the stria medullaris thalami. This begins in the region of the anterior perforated substance, where it picks up olfactory impulses. It ascends to the lateral border of the roof of the third ventricle which conducts it to the anterior end of the habenular trigone. Commissural fibres connect the right and left habenular nuclei to form the habenular commissure. The efferent fibres of the epithalamus form the slender fasciculus retroflex that descends into the tegmentum of the midbrain where motor responses may be excited via the reticulospinal tracts.

7. The Cranial Nerves

IN THE PROCESS of tracing the ascending and descending pathways of the brain, we have located the grey matter that contains the sensory and motor nuclei. In this chapter we will describe the nuclei in these cell masses and the nerves connected with them.

There are twelve pairs of cranial nerves. Each one has a number and a name. The numbers indicate the order of attachment to the brain; the names indicate their function or some anatomical characteristic.

I. THE FUNCTIONAL COMPONENTS OF EACH CRANIAL NERVE

NERVE 1: THE OLFACTORY NERVE

This is the nerve of the sense organ for smell. It has several unique features. First, its ganglion cells are in the epithelium of the nose not in a cluster near the brain. Second, its nerve fibres are grouped into many small bundles that enter the brain (the olfactory bulb) like the bristles of a brush, and third, its sensory nucleus is in the olfactory bulb not in the brain stem.

NERVE 2: THE OPTIC NERVE

This nerve comes from the retina of the eye. It is not really a nerve but a fibre tract because the eye developmentally is a part of the brain. The retina has three layers of nerve cells. One layer contains the cells that correspond to the ganglion cells of a typical nerve. The other two are relay stations on the pathway to the diencephalon. The fibres of the third layer of cells are the ones that make up the bundle called the optic nerve.

NERVE 3: THE OCULOMOTOR NERVE

This nerve gets its name because it is a motor nerve that supplies most of the muscles of the eye. It supplies striated muscles derived from somite mesoderm and also some smooth muscles—the constrictor of the pupil and the ciliary muscle.

NERVE 4: THE TROCHLEAR NERVE

This is a motor nerve that supplies one striated muscle of the eye that is derived from somite mesoderm. The tendon of this muscle plays over a pulley (trochlea).

NERVE 5: THE TRIGEMINAL NERVE

This nerve has many of the characteristics of a spinal nerve. Its sensory fibres are from sense organs of pain, temperature, touch, and position, and enter the brain in a separate root. Its motor fibres supply striated muscles and leave the brain in a bundle that corresponds to a ventral root. The motor fibres differ developmentally from those of spinal nerves, however, because they supply muscles that are derived from the mesoderm of the first branchial arch (muscles of mastication) not from the mesoderm of a somite.

The nerve gets its name (*trigeminal*, three buds) because it has three large branches, one to the eye region (sensory), one to the upper jaw region (sensory), and one to the lower jaw region (sensory and motor). Its sensory ganglion is the semilunar (Gasserian) ganglion.

NERVE 6: THE ABDUCENS NERVE

This, like the trochlear, is a motor nerve that supplies a striated muscle derived from the mesoderm of a somite. It supplies the muscle that turns the eye outwards—abducts it.

NERVE 7: THE FACIAL NERVE

This nerve contains four kinds of fibres. (1) Motor fibres to striated muscles that are derived from the mesoderm of the second branchial arch. Most of these muscles are attached to the skin of the head and neck and are known as the muscles of expression. (2) Visceral motor (preganglionic) fibres that carry impulses destined for the lacrimal, nasal, palatal, and salivary glands. (3) Sensory fibres from sense organs of pain and possibly temperature and touch in some areas of the skin and mucous membrane that are supplied by the trigeminal nerve. These areas of overlapping nerve supply are on the tongue, palate, in the external auditory meatus, and on the outer aspect of the ear drum. (4) Visceral sensory fibres that carry impulses from sense organs of taste on the anterior-two thirds of the tongue.

The nerve gets its name because its large motor branches fan out over the face. Its sensory ganglion is the geniculate ganglion.

NERVE 8: THE VESTIBULOCOCHLEAR NERVE

This nerve has two kinds of sensory fibres in two bundles known as the cochlear (auditory) and the vestibular nerves. The vestibular nerve comes from special sense organs that detect movement of the head and record its position relative to the pull of gravity. The sensory ganglia of the two nerves are within the temporal bone in the vestibule and the cochlea.

NERVE 9: THE GLOSSOPHARYNGEAL NERVE

This, like the facial nerve, contains four kinds of fibres. (1) Motor fibres to a striated muscle of the pharynx (stylopharyngeus) that is derived from the mesoderm of the third branchial arch. (2) Visceral motor (preganglionic) fibres that carry impulses destined for the parotid and also the other salivary glands. (3) Sensory fibres from sense organs of pain, temperature, and touch, in the wall of the pharynx, in the posterior third of the tongue, and from the middle ear. Stimulating these sensory endings in the pharynx initiates swallowing. (4) Visceral sensory fibres that carry impulses from sense organs of taste in the posterior third of the tongue, from chemo-receptors and pressure receptors in the carotid sinus (for the regulation of respiration and blood pressure).

The glossopharyngeal nerve gets its name from its chief areas of distribution, the tongue and the pharynx. The ganglion cells of the sensory part of the nerve are in one or other of two clusters that are located in the portion of the nerve in the jugular foramen.

NERVE 10: THE VAGUS NERVE

This also has four kinds of fibres. (1) Motor fibres to the striated muscles of the larynx and pharynx derived from the mesoderm of the fourth and fifth branchial arches. (2) Visceral motor (preganglionic) fibres that carry impulses destined for the thoracic and abdominal viscera as far as the left colic flexure. (3) Sensory fibres from sense organs of pain and possibly temperature and touch in the skin of the external auditory meatus. This is an area of skin also supplied by the trigeminal and the facial nerves. (4) Visceral sensory fibres that carry impulses from sense organs of taste (epiglottis); from stretch receptors in the wall of the heart, superior vena cava, aorta, and the bifurcation of the common carotid, all of which regulate blood pressure and heart rate; from stretch receptors in the lung, which regulate the rate and depth of respiration.

The vagus nerve gets its name from its long meandering course (*vagus*, wandering). The ganglion cells of the sensory part of the nerve are in one or other of two clusters that are located in the portion of the nerve in the jugular foramen.

NERVE 11: THE ACCESSORY NERVE

This is a motor nerve that supplies striated muscles that are derived from the mesoderm of the most caudal of the branchial arches. Some of these muscles are in the wall of the pharynx and larynx, but two are superficial muscles of the neck, the sternomastoid and the trapezius.

NERVE 12: THE HYPOGLOSSAL NERVE

This is a motor nerve that supplies striated muscles derived from the mesoderm of somites. It supplies the muscles of the tongue.

II. CLASSIFICATION OF CRANIAL NERVE NUCLEI

A. SENSORY NUCLEI

As the sensory fibres of nerves 5, 7, 8, 9, and 10 enter the brain they sort themselves out according to function. All pain fibres go into one bundle, all visceral fibres into another, and so on. Each of these bundles then proceeds to its own relay station. It is these relay stations for sensory nerve fibres that are called sensory nuclei and each one is given the name of its afferent fibre bundle.

B. MOTOR NUCLEI

The motor nuclei are clusters of cells whose fibres leave the central nervous system in a nerve and carry impulses to effectors. Nerve fibres may end in striated muscles derived from the mesoderm of a somite, or in striated muscles derived from the mesoderm of a branchial arch, or they may end in an autonomic ganglion. Those that end in an autonomic ganglion synapse with cells that relay the impulses to smooth muscles, heart, or glands. Hence motor nuclei are classified as somite, branchial, or preganglionic nuclei. The cells of an autonomic ganglion are known as postganglionic neurons. The cells of the somite and branchial motor nuclei look alike, they are very large multipolar cells. The cells of the preganglionic nuclei are multipolar also but they are very small.

III. THE LOCATION OF SENSORY AND MOTOR NUCLEI IN A CROSS-SECTION OF EACH OF THE SEGMENTS OF THE CENTRAL NERVOUS SYSTEM

In a section of the spinal cord (figure 59) the sensory nuclei are in the dorsal horn, the motor nuclei in the ventral horn. Serial sections of

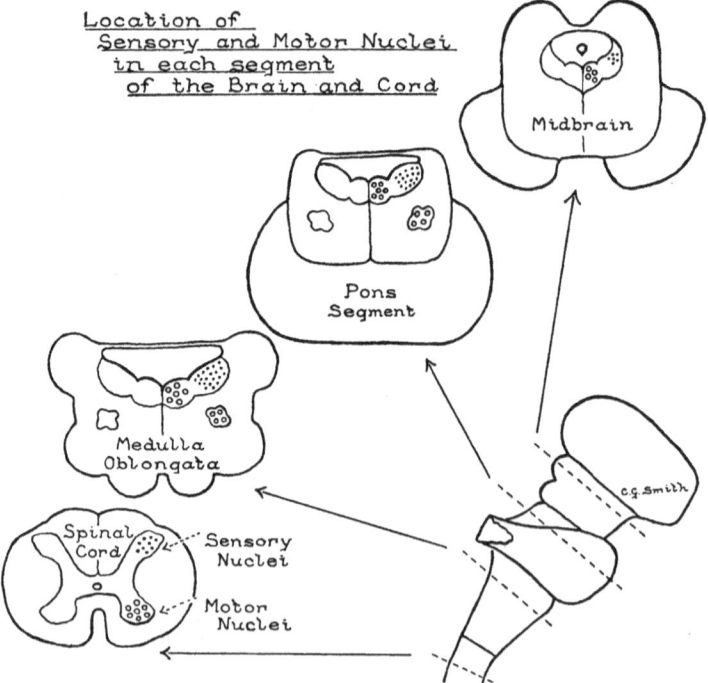

FIGURE 59. The location of the sensory and motor nuclei in cross-sections of the central nervous system.

the transitional zone between the cord and the medulla reveal that the irregularly rectangular band of grey matter formed by the dorsal and ventral horns extends up into the brain. However, it rotates and shifts dorsally in the medulla to lie in the floor of the fourth ventricle. The motor nuclei (ventral horn grey matter) lie in the medial part of the floor and the sensory nuclei (dorsal horn grey matter) lie in the lateral part of the floor. Serial sections also reveal that the grey matter of the ventral horn divides as it extends into the medulla and a slender column of cells remains in a ventral location in line with the ventral horn of the cord. This slender column of grey matter contains the motor nuclei of muscles derived from branchial arch mesoderm. The motor nuclei in the floor of the ventricle supply striated muscles also but these are derived from the mesoderm of somites. Along with these there are motor nuclei that send preganglionic fibres to autonomic ganglia.

In a section of the tegmentum of the pons the locations of the motor and sensory nuclei are the same as in the medulla.

In the midbrain the column of grey matter that contains the branchial motor nuclei has dropped out. However, the grey matter containing somite motor nuclei, preganglionic motor nuclei, and sensory nuclei is still present. It lies ventral to the aqueduct and forms a part of the central grey matter.

IV. THE MOTOR NUCLEI OF THE BRAIN STEM

The three kinds of motor nuclei—somite, branchial, and preganglionic (visceral)—are in three columns like strings of irregularly spaced beads (figure 60). The column of branchial motor nuclei reaches as far up as the middle of the pons. The other two reach the upper border of the midbrain. Each of the nuclei is cylindrical and may be long or short.

The three columns are together at the caudal end of the medulla in the grey matter of the ventral horn. The branchial column ascends in the central part of the reticular substance of the medulla and pons but the other two shift to the midline and dorsally to lie in the floor of the fourth ventricle in the hindbrain and ventral to the aqueduct in the midbrain.

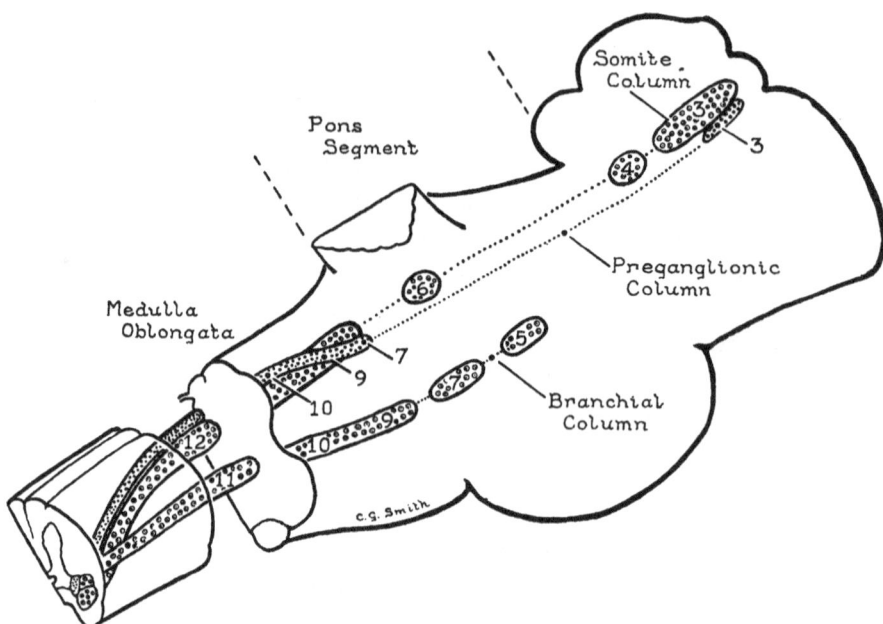

FIGURE 60. The columns of motor nuclei projected onto the lateral aspect of the brain stem.

A. THE COLUMN OF SOMITE MOTOR NUCLEI

The motor nuclei of this column are nuclei of the twelfth (hypoglossal), sixth (abducens), fourth (trochlear), and third (oculomotor) nerves (figure 61).

1. THE NUCLEUS OF THE HYPOGLOSSAL NERVE

This is almost as long as the medulla. It produces a ridge, called the hypoglossal trigone, in the floor of the ventricle next to the midline. This ridge has a pointed caudal end because the preganglionic nucleus of the vagus nerve is dorsal to it in the closed medulla and gradually shifts laterally in the open medulla to leave it uncovered.

The nerve fibres of this nucleus course ventrally just lateral to the medial lemniscus to emerge as a series of rootlets between the pyramid and the olive.

2. THE NUCLEUS OF THE ABDUCENS NERVE

This is a small cluster of cells. It is in the caudal part of the pons segment and elevates the floor of the ventricle between the sulcus

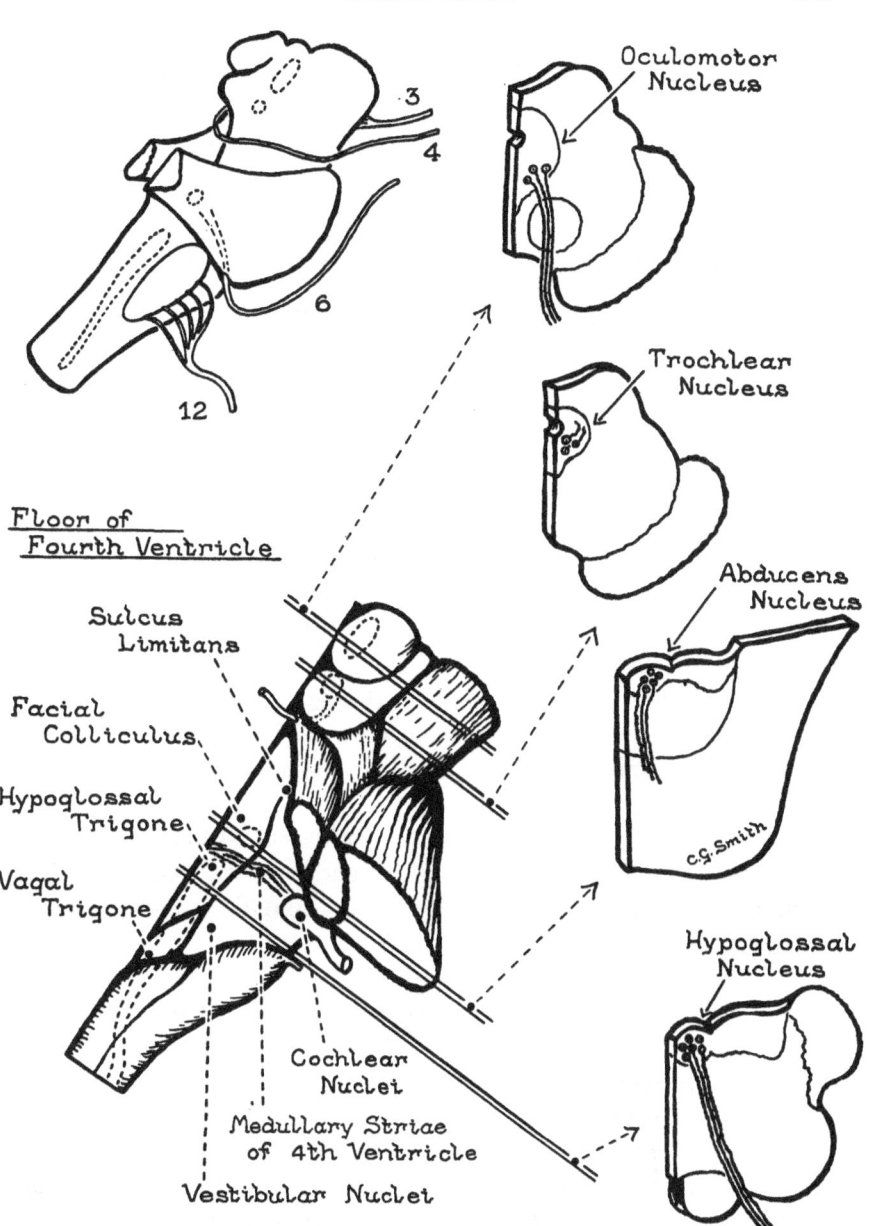

FIGURE 61. The motor nuclei supplying muscles derived from somites.

limitans and the midline. The mound-like elevation is called the facial colliculus because the facial nerve helps to produce it, as we shall see.

The nerve fibres of this nucleus course ventrally and caudally to avoid the thickest part of the pons and emerge between it and the pyramid.

3. THE NUCLEUS OF THE TROCHLEAR NERVE

This too is a small cluster of cells. It is located in the ventral part of the central grey matter of the midbrain at the level of the inferior colliculus.

The nerve fibres of this nucleus have a peculiar course. They course dorsally and caudally to cross the midline in the superior velum. Thus the trochlear nerve has its nucleus on the opposite side of the brain.

4. THE NUCLEUS OF THE OCULOMOTOR NERVE

This is a cylindrical cluster of cells that extends from the trochlear nucleus to the upper border of the midbrain. It lies close to the midline in the ventral part of the central grey matter. Its cells are in five groups, one for each muscle it supplies.

The nerve fibres course ventrally through the red nucleus and emerge at the medial border of the crus cerebri. The fibres to the superior rectus muscle come from cells in the nucleus of the opposite side.

B. THE COLUMN OF PREGANGLIONIC NUCLEI

The motor nuclei of this column are nuclei of the tenth (vagus), ninth (glossopharyngeal), seventh (facial), and third (oculomotor) nerves (figure 62).

1. THE NUCLEI OF THE VAGUS, GLOSSOPHARYNGEAL, AND FACIAL NERVES

These three nuclei form an uninterrupted cell column that extends the length of the medulla. In the caudal part of the medulla this column is at the midline dorsal to the hypoglossal nucleus. As it enters the open part of the medulla it is visible in the floor of the ventricle and forms the ridge known as the vagal trigone or ala cinerea. This ridge grows narrower, tapering to a point toward the pons, because the cell column gradually shifts laterally to lie lateral to the hypoglossal nucleus and ventral to the vestibular nucleus.

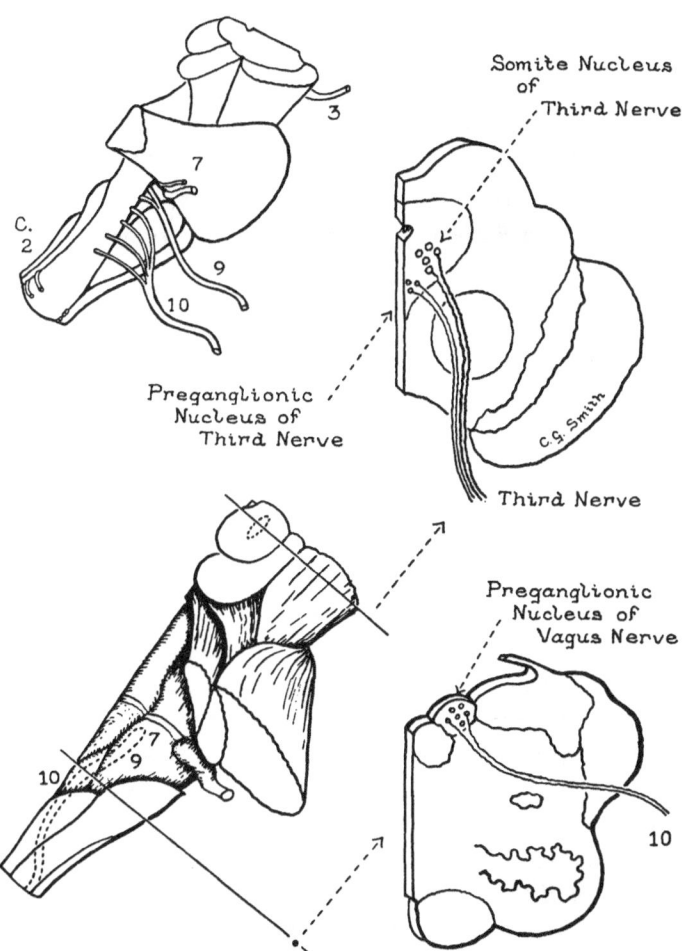

FIGURE 62. The preganglionic or visceral motor nuclei.

A small rostral segment of this column sends its fibres into the facial nerve and is known as the superior salivary nucleus. The middle segment sends its fibres into the glossopharyngeal nerve and is known as the inferior salivary nucleus. The caudal and the longest portion sends its fibres into the vagus nerve and is called the dorsal motor nucleus of the vagus.

The nerve fibres from each of these nuclei course laterally to emerge in thread-like rootlets along the dorsal border of the olive.

2. THE NUCLEUS OF THE OCULOMOTOR NERVE

This nucleus is a short, cylindrical cluster of cells that lies medial to the upper end of the somite motor nucleus of the oculomotor nerve. It is popularly known as the nucleus of Edinger-Westphal.

The nerve fibres of this nucleus join the other fibres of the third nerve and course with them in one bundle as far as the orbit. There they leave the other fibres to enter the ciliary ganglion.

C. THE COLUMN OF BRANCHIAL MOTOR NUCLEI

The motor nuclei of this column are nuclei of the eleventh (accessory), tenth (vagus), ninth (glossopharyngeal), seventh (facial), and fifth (trigeminal) nerves (figure 63).

1. THE NUCLEI OF THE ACCESSORY, THE VAGUS, AND GLOSSOPHARYNGEAL NERVES

These three nuclei form an uninterrupted column that extends the length of the medulla in the central part of the reticular substance. It is known as the nucleus ambiguus because its cells are widely spaced and its borders are irregular. The cells of the glossopharyngeal nucleus are in a small rostral segment of the nucleus ambiguus. Caudal to this is a long segment containing the cells of the vagus nerve fibres and caudal to this in turn is the short segment for the accessory nerve. If the nucleus ambiguus is traced toward the cord in serial sections it is found to be continuous with a column of cells in the ventral horn that extends to the fifth cervical segment.

The nerve fibres of the cells at the rostral pole of the nucleus are joined by fibres of the inferior salivary nucleus and course laterally to emerge behind the olive in two or three rootlets that unite to form the ninth nerve. Similarly the nerve fibres of the vagal portion of the nucleus ambiguus are joined by preganglionic fibres of the dorsal motor nucleus of the vagus and emerge behind the olive as three or four rootlets which unite to form the vagus nerve.

The nerve fibres of the cell column in the cord that is continuous with the nucleus ambiguus leave the ventral horn and course laterally through the lateral funiculus to emerge in groups along a longitudinal line that is nearer the dorsal than the ventral roots. These intermediate roots ascend alongside the cord and join to form the spinal part of the accessory nerve. This nerve ascends in the subarachnoid space and enters the skull through the foramen magnum. Inside the skull it picks

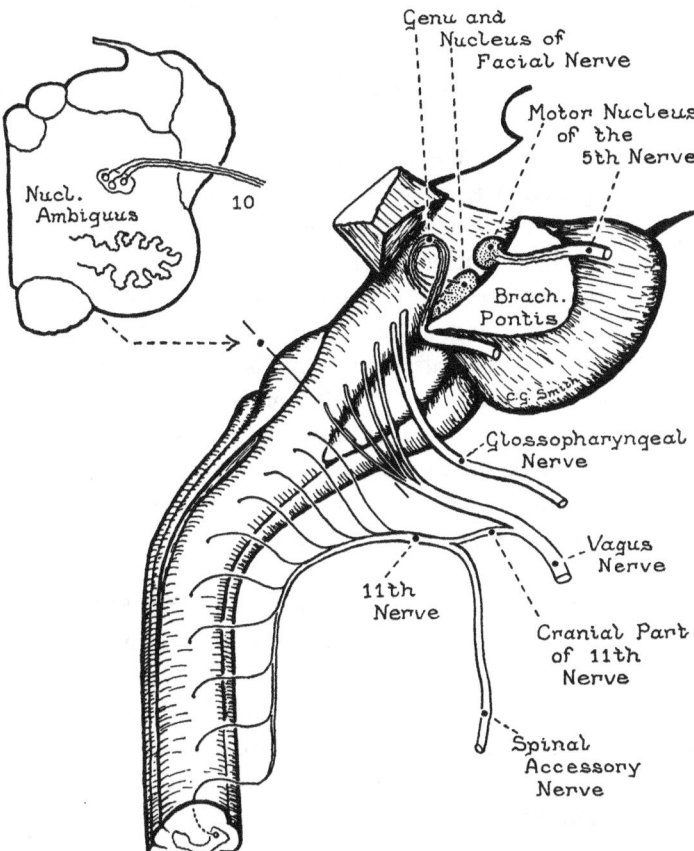

FIGURE 63. The motor nuclei supplying muscles derived from the mesoderm of the branchial arches.

up a few rootlets from the caudal part of the nucleus ambiguus. The nerve, having obtained all its fibres leaves the skull again through the jugular foramen. In this foramen, its cranial and spinal fibres separate. The spinal fibres form an external branch that supplies the sternomastoid and the trapezius muscles. The cranial fibres join the vagus nerve and are distributed through its branches to help supply the striated muscles of the larynx and pharynx. The fibres of the nucleus ambiguus that form the cranial division of the accessory nerve are essentially vagal fibres that delay joining their fellows until the vagus nerve enters the jugular foramen.

2. THE NUCLEUS OF THE FACIAL NERVE

This nucleus is a fusiform cluster of cells 2 or 3 mm. long, located in the caudal part of the pons just dorsal to the superior olive (figure 35).

The fibres of this nucleus have a very unusual course. Instead of passing laterally and caudally a distance of 2 mm. to the caudal border of the pons, where the facial nerve emerges, it takes a circuitous route around the abducens nucleus (figure 63). The course of the facial nerve within the pons reveals, like a vapour trail, the course of migration of its nucleus during development. There is no satisfactory explanation for it. As it emerges at the caudal border of the pons it is almost in line with the ninth and tenth nerve rootlets. It is just ventral to the attachment of the acoustic nerve separated from it only by a thread-like filament known as the nervus intermedius. This is looked upon as the sensory root of the facial nerve but it contains preganglionic fibres as well as sensory fibres.

3. THE NUCLEUS OF THE TRIGEMINAL NERVE

This nucleus is about 2 mm. long and is located in the middle of the pons segment in line with the facial nucleus.

The nerve fibres pass directly laterally in a bundle that penetrates the brachium pontis to emerge in line with the seventh, ninth, and tenth nerves. It is the motor root of the trigeminal nerve but it is not ventral to the sensory root as might be expected.

IV. THE SENSORY NUCLEI OF THE BRAIN STEM

Let us begin our study of the sensory nuclei by identifying those that can be seen in a section of the open medulla (figure 64). In such a section the rectangular mass of sensory nuclei in the floor of the ventricle is divided into a dorsal and a ventral layer. The dorsal layer is the sensory nucleus of the vestibular nerve. The ventral layer is divided into a small medial and a large lateral part by a small bundle of fibres. The medial nucleus is that of the tractus solitarius (visceral sensory) and the lateral nucleus is that of the spinal tract of the fifth nerve. If we turn now to the diagram in figure 65, in which these nuclei are represented as if they were projected onto the lateral surface

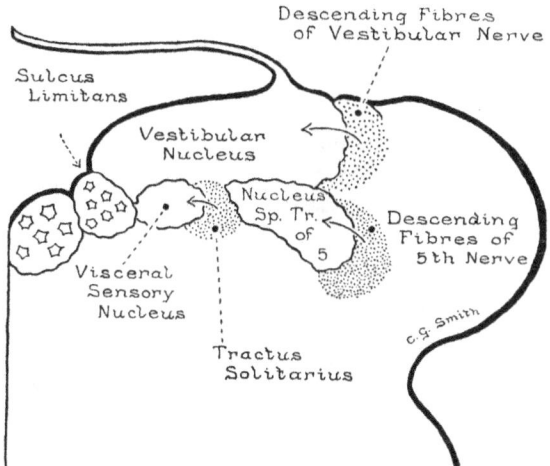

FIGURE 64. A cross-section of the open medulla to show the subdivisions of the sensory column of grey matter (the dorsal horn of the cord).

and in which the plane of the section of the open medulla is shown, we can see how far these nuclei extend upwards and downwards in the brain stem. Beginning with the nucleus of the spinal tract of the fifth nerve we see that it becomes very large in the caudal part of the medulla and extends to the spinal cord where it forms the crest of the dorsal horn of the first cervical segment. This portion of this columnar nucleus receives the pain and temperature fibres of all the cranial nerves. If this same column is followed up into the pons it is seen to enlarge there also in the middle part of this segment to form the main (superior) sensory nucleus of the fifth nerve. It is the relay station for the touch fibres of all the cranial nerves. Toward the midbrain this nucleus is continuous with a slender column that extends up the length of the midbrain. This is the mesencephalic nucleus of the fifth nerve and is the relay station for fibres of all cranial nerves that carry impulses from sense organs in muscles (position sense).

Turning now to the nucleus of the tractus solitarius (visceral sensory) we see that it extends caudally into the lateral part of the dorsal horn of the cord. Traced upwards it ends at the junction of medulla and pons. This nucleus receives the visceral sensory fibres of the seventh, ninth, and tenth nerves. These penetrate the medulla to its lateral side and turn caudally there to form the tractus solitarius.

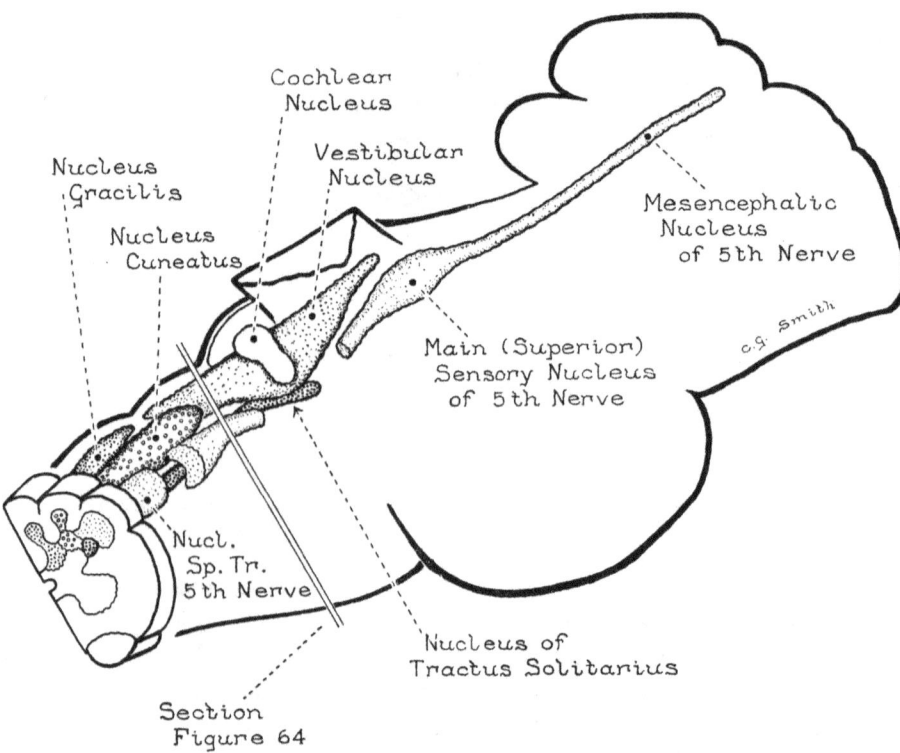

FIGURE 65. The subdivisions of the sensory column of grey matter projected onto the lateral aspect of the brain stem.

The vestibular nucleus forms the width of the floor of the ventricle lateral to the sulcus limitans and extends laterally slightly beyond it (figure 64). This portion of the floor of the ventricle, and with it the vestibular nucleus, tapers to a point (see figure 61) as the ventricle grows narrower rostrally and caudally. If we examine a section of the open medulla at the level of the lateral recess (see figure 68) we see that the grey matter of the floor of the ventricle extends out onto its floor. This lateral extension of the sensory grey matter is the nucleus of the cochlear division of the eighth nerve. It rests on the restiform body.

The remaining subdivisions of the sensory grey matter of the brain stem are the nuclei gracilis and cuneatus. These are displaced sensory nuclei of the spinal cord because they receive the ascending branches of spinal sensory nerve fibres of touch and the sense of position. Each of these nuclei is a column of cells that begins close to the first cervical segment and extends to the vestibular nucleus in the core of its afferent fibre bundle. In sections these nuclei can be seen to be parts of the cranial prolongation of the grey matter of the dorsal horn.

V. THE AFFERENTS AND EFFERENTS OF THE THREE PARTS OF THE SENSORY NUCLEUS OF THE FIFTH NERVE

The afferent fibres of the sensory nuclei of the fifth nerve come from sense organs that excite pain, warmth, cold, touch, and a sense of position (figure 66). These fibres reach the brain in the fifth, seventh, ninth, and the tenth cranial nerves but most of them arrive in the fifth nerve. For this reason the fibre bundles formed by these fibres in the brain are called tracts of the fifth nerve and the nuclei in which they end are named accordingly.

The fifth nerve penetrates the middle of the side of the pons and reaches the tegmentum at the side of the main sensory nucleus of the fifth nerve. Here most of the touch fibres end. The pain and the temperature fibres plus a few touch fibres turn and descend lateral to this nucleus as the spinal tract of the fifth nerve. The touch fibres end in the upper part of the sensory nucleus of this tract but the pain and temperature fibres descend farther to end partly in the lower part of the medulla and partly in the first cervical segment of the cord.

The sensory fibres from receptors in muscles (sense of position) part company from the touch and pain fibres at the side of the main sensory nucleus and course upwards along the lateral side of the mesencephalic nucleus of the fifth nerve to end in it.

The general sensory fibres of the seventh, ninth, and tenth nerves leave these nerves as they enter the medulla and ascend or descend in the tract of fibres formed by the fifth nerve to reach their appropriate relay station.

The efferent fibres of the nuclei of the fifth nerve that ascend toward the cerebral hemisphere cross the midline and join the appropriate sensory pathway ascending from the spinal cord. Those from the nucleus in the medulla and first cervical segment join the lateral spinothalamic tract, while those from the nuclei in the pons and midbrain join the medial lemniscus.

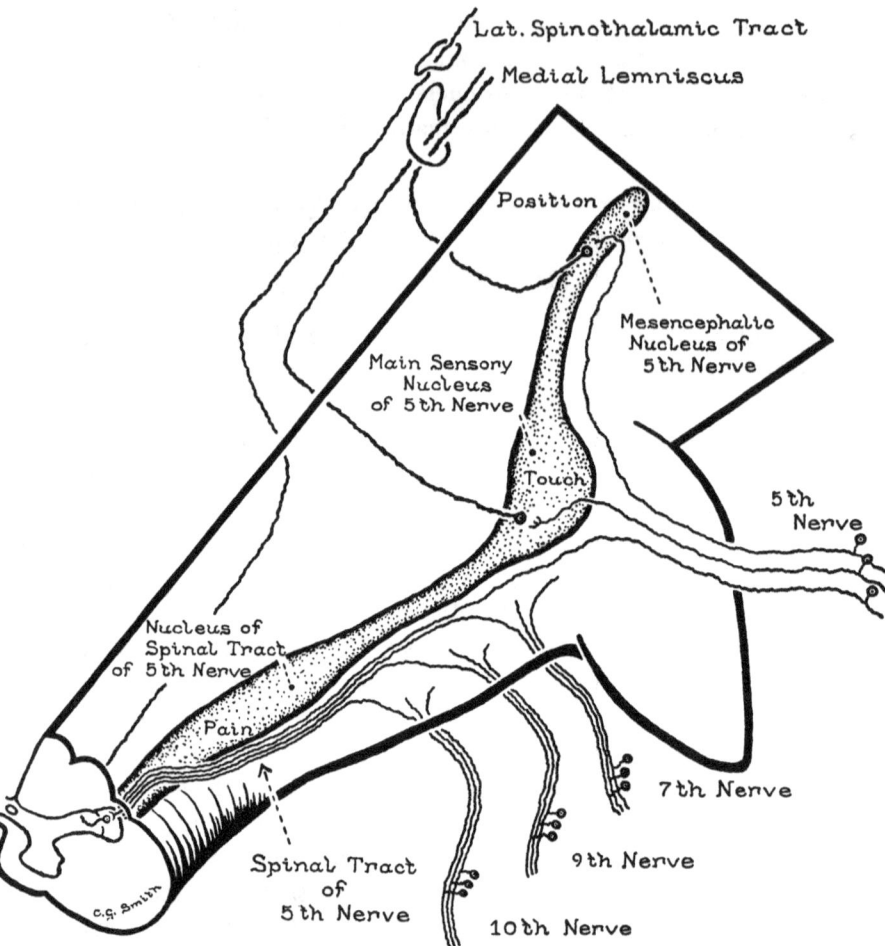

FIGURE 66. The column of grey matter made up of general sensory nuclei (pain, temperature, touch and sense of position) showing its afferent fibres and the efferent fibres that ascend to the forebrain.

VI. THE AFFERENTS AND EFFERENTS OF THE NUCLEUS OF THE TRACTUS SOLITARIUS

The afferents of the nucleus of the tractus solitarius come from taste buds, heart, aorta, lungs, and alimentary tract and enter the brain in nerves 7, 9, and 10 (figure 67). These fibres course medially to the side of their nucleus and descend there to form the tractus solitarius. This bundle is surrounded by grey matter and therefore stands out clearly in a section stained to show the fibres. It gets smaller caudally as fibres drop out to end in the nucleus.

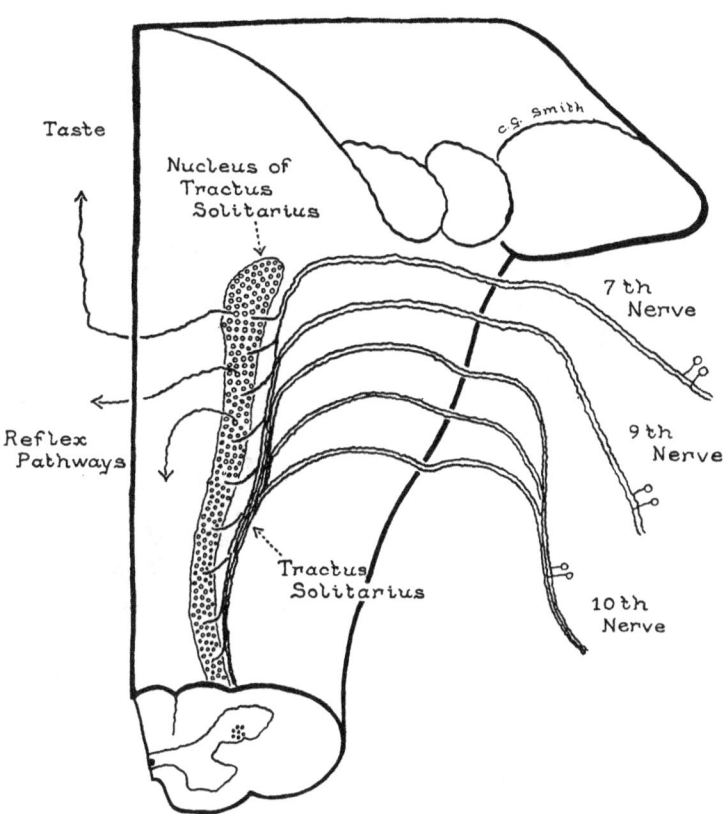

FIGURE 67. The column of grey matter made up of visceral sensory nuclei—the nucleus of tractus solitarius—showing its afferent and efferent fibres.

The course of the fibres of this nucleus that ascend to excite a sensation of taste is not known. Maybe they are in the medial lemniscus. Some efferent fibres enter the reticular substance through which impulses are conveyed to the appropriate motor nuclei to initiate such reflex movements as swallowing, or grimacing in response to a sour taste.

VII. THE AFFERENTS AND EFFERENTS OF THE VESTIBULAR NUCLEI

The vestibular nerve and the cochlear nerve embrace the restiform body at the caudal border of the brachium pontis (figures 68, 69). Here the vestibular nerve goes ventral to the restiform body to reach

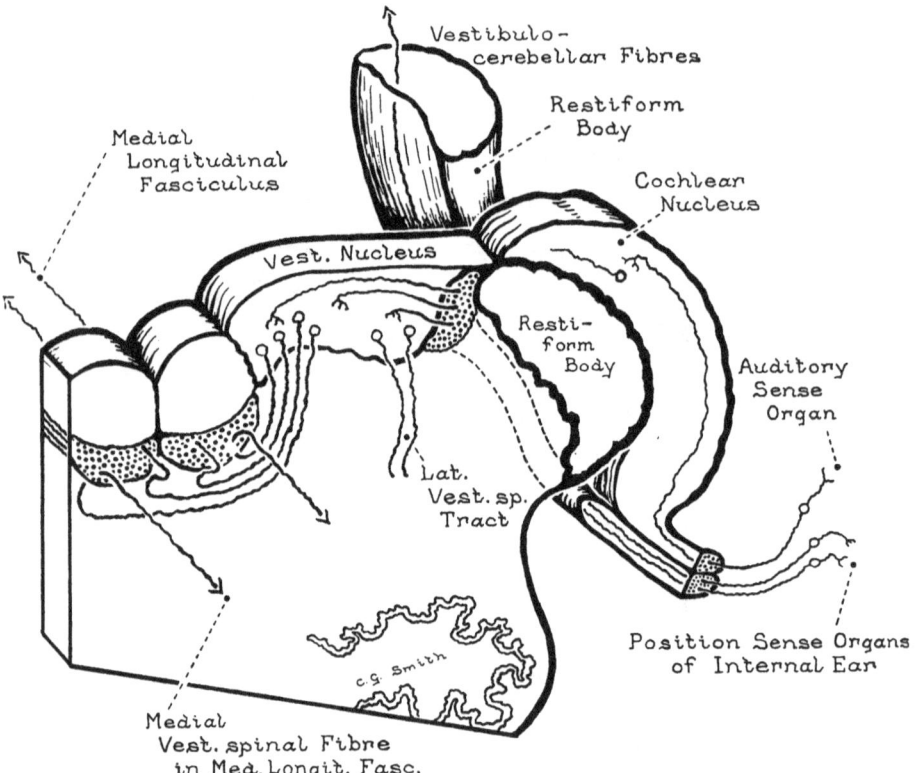

FIGURE 68. The caudal aspect of a thick slice of the open medulla to show the formation of the medial longitudinal fasciculus and the vestibulospinal tracts.

the lateral border of the vestibular nucleus. Its fibres ascend and descend (figure 69) along the lateral side of the vestibular nucleus and then turn inwards to synapse with its cell bodies. Some of the ascending fibres join the deep surface of the restiform body and go directly to the cerebellum. These fibres are going to the oldest part of the cerebellum.

The known connections of the vestibular nuclei are all reflex connections. Ascending pathways that excite the sensation of movement have not been traced. The reflex connections are, however, well known and most important. These initiate reflex adjustments in posture of the eyes, head, trunk, and limbs in response to (a) changes in the position of the head or (b) movements of the head. These postural adjustments always involve the two sides of the body, hence the reflex pathways

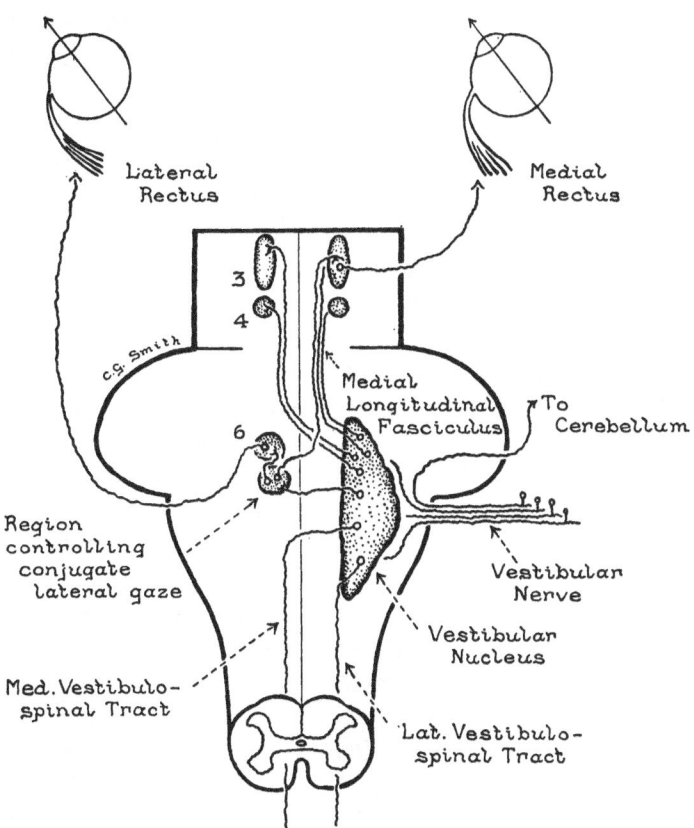

FIGURE 69. The dorsal aspect of the brain stem to show the connections of the right vestibular nerve and the pathways leaving the right vestibular nucleus.

are paired. In some cases they will excite the same movement on both sides, in other cases the right and left pathways will act, functionally, like the two ropes of a rudder and cause contraction of the muscles of one side and inhibition of the contraction on the other.

The reflex pathways can be divided into ascending and descending groups. The ascending pathways excite postural adjustments of the eyes. These postural adjustments can be resolved to two planes of movement, (*a*) a vertical, up and down and (*b*) horizontal, to the right and left. Up and down movements are controlled by muscles supplied by the third and fourth nerves which have their nuclei in the midbrain. Since the two eyes always move together, it follows that fibres from the vestibular nuclei of one side ascend on both sides.

These ascending fibres lie next to the midline ventral to the column of somite motor nuclei in a compact bundle called the medial longitudinal fasciculus.

Movements of the eyes to the right and left are controlled by the medial and the lateral rectus muscles which are supplied by the third and the sixth nerves respectively. Thus when the eyes look to the left, the left lateral rectus is excited via the sixth nerve, and the right medial rectus is excited via the third nerve.

How are the impulses conveyed to the nuclei of these two motor nerves located on opposite sides of the brain? The pathway involved in the response when the head is rotated to the right is shown in figure 69. If, when seated, the chair is rotated to the right the eyes reflexly move to the left as if to retain the visual field. The response as far as the vestibular system is concerned is due to the excitation of the sense organs of the right internal ear. Impulses set up in the right internal ear are conveyed to the cells of the right vestibular nucleus which sends fibres across the midline to a group of cells not yet identified anatomically but known to be close to, if not within, the nucleus of the sixth nerve. Some of these cells send fibres to the adjacent nucleus of the sixth nerve, others send fibres to the remote nucleus of the right third nerve. To reach the right oculomotor nucleus, the fibres cross the midline and ascend in the medial longitudinal fasciculus. Thus a so-called centre for conjugate lateral gaze (both eyes together) is present in the pons segment close to the nucleus of the sixth nerve. All horizontal movements of the eyes whether reflex or voluntary are activated through this centre.

We turn now to the descending pathways that reach the motor nuclei of the spinal nerves. Here again there are a pair of pathways, an uncrossed and a crossed. The uncrossed, called the lateral vestibulospinal tract, is the most publicized because cutting it in animals abolished decerebrate rigidity. Its fibres descend dorsal to the olive and in the spinal cord they are in the ventral funiculus. The fibres do not form a compact bundle. They synapse with the cells of the ventral horn. The crossed pathway is made up of fibres that cross the midline to form the descending limb of the medial longitudinal fasciculus. This bundle is next to the midline and ventral to the nucleus of the twelfth nerve, until it reaches the cord where it lies in the ventral funiculus. Its fibres end in the ventral horn. This crossed pathway unlike the uncrossed one is inhibitory not excitatory.

Thus excitation of the right vestibular nerve will lead to increased extensor tone in muscles on its own side and depression of muscle tone in muscles of the opposite side. For this reason a subject tends to fall to one side, the inhibited side, on trying to stand after a period of rotation.

VIII. THE AFFERENTS AND EFFERENTS OF THE COCHLEAR NUCLEI

The cochlear division of the acoustic nerve passes to the dorsal surface of the restiform body in the floor of the lateral recess. Here it encounters the lateral extremity of the cochlear nucleus. The cells of this nucleus send their fibres around the restiform body to pass through the reticular substance to the other side where after passing through the medial lemniscus or around it they ascend in a bundle known as the lateral lemniscus. A small number of fibres that leave the cochlear nucleus do not cross the midline. They synapse with cells in the superior olive which serves as a relay station on a homolateral pathway to the forebrain. The fibres ascending from the superior olive join the lateral lemniscus.

IX. A SUMMARY OF THE RELATIONSHIPS OF THE SENSORY AND MOTOR NUCLEI OF THE MEDULLA

Just above the decussation of the pyramids four sensory and three motor nuclei are present in the cranial extensions of the grey matter of the dorsal and ventral horns respectively (figure 70C). At the level of the caudal part of the fourth ventricle (figure 70B) two of the motor nuclei, the hypoglossal and the preganglionic nucleus of the vagus, and the four sensory nuclei form a chain that extends from the midline dorsally to the floor of the fourth ventricle and then laterally close to the dorsal surface of the medulla. The third motor nucleus, the nucleus ambiguus, is detached from the rest and lies in the core of the reticular substance. At a slightly higher level (figure 70A) the sensory nucleus of the tractus solitarius and the motor preganglionic nucleus of the glossopharyngeal nerve, which replaces the preganglionic nucleus of the vagus, are crowded laterally by the hypoglossal nucleus and out of line. These two displaced nuclei lie lateral to the sulcus limitans and ventral to the vestibular nucleus

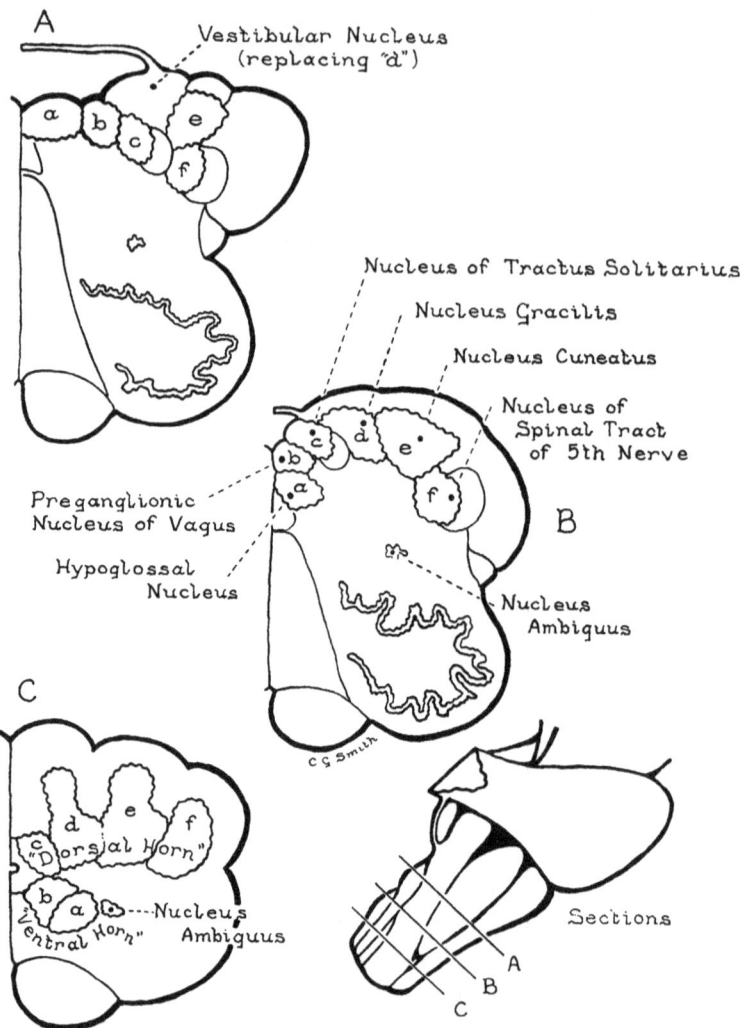

FIGURE 70. Serial sections of the medulla to show the relationships of the subdivisions of the grey matter of the spinal cord to the cranial nerve nuclei.

which has replaced the nucleus gracilis. Note that the vestibular nucleus is partly in the floor of the ventricle and partly lateral to it on the dorsal surface. At a higher level in the region of the lateral recess of the ventricle (figure 68) the vestibular nucleus enlarges laterally and replaces the nucleus cuneatus as well as the nucleus gracilis.

8. The Reticular Formation

I. STRUCTURAL CHARACTERISTICS

THE CENTRAL NERVOUS SYSTEM is made up of grey and white matter and reticular substance. Grey matter is made up of closely packed cell bodies, white matter of closely packed nerve fibres, and the reticular substance is a mixture of cell bodies and nerve fibres. The name "reticular substance" is fitting because the nerve fibres are loosely arranged and in a section they form a net-like or web-like structure, a *reticulum*. The spaces in the meshwork are occupied by cell bodies. In any section of the brain stem the reticular substance forms a central core. It may be likened to padding which fills the spaces that are not filled by compact nuclei and fibre tracts. The reticular substance converts the skeleton-like structure made up of nuclei and fibre tracts into a solid mass.

II. FUNCTIONAL CHARACTERISTICS AND AFFERENTS AND EFFERENTS

The reticular substance is a pool of neurons within which occurs the correlation of the sensory data discharged into the sensory nuclei by the spinal and cranial nerves. Some examples of its functioning are the complex reflexes of swallowing, sneezing, coughing, and the regulation of respiration and circulation. The fibres of the reticular substance of necessity form a loosely woven network because the afferent fibres come from so many different sources and the efferents, similarly, have many widely separated destinations.

Recent study has shown that the reticular substance in addition to its basic role as a correlative mechanism for complex reflexes, is organized to influence the activity of the whole nervous system. To this end it gives rise to ascending and descending pathways. The ascending pathways have their cell bodies in the reticular formation of the medulla, pons, and midbrain. They have an activating influence on the cerebral hemisphere and induce a state of alertness. The fibres of the ascending pathways are diffusely disposed within the reticular substance of the medulla and pons. In the midbrain they begin to crowd together into a bundle called the **reticulothalamic tract.** This tract ascends into the diencephalon. Destruction of this fibre bundle in a monkey causes it to go into a state resembling normal sleep that is not temporary but continues as long as the animal lives. The monkey can be roused by strong stimuli, but it lapses into unconsciousness again with the cessation of the stimulus.

In addition to the pathways that ascend to the forebrain, the reticular formation has descending pathways called the **reticulospinal tracts.** The fibres of these tracts have their cell bodies in the reticular substance of the medulla and the pons and descend diffusely to enter the ventral funiculus and the ventral part of the lateral funiculus of the cord. The pathways from the reticular formation of one side descend into both right and left sides of the spinal cord. They end on the ventral horn cells, and carry impulses that regulate respiration and circulation and play an accessory role in motor responses. Stimulation of the reticular formation can dramatically facilitate or inhibit movements that are excited reflexly or by the stimulation of other parts of the nervous system. In general, inhibitor responses are obtained from the reticular formation of the medulla, facilitatory responses from the reticular formation of the pons. The response is not predictable because it is influenced by what is happening at the moment in other parts of the body. The reticulospinal pathways are also believed to be the motor pathway through which the patterns of movement associated with emotional responses, such as laughing and rage, are initiated.

The role of the reticular formation in the control of movement suggests it should have connections with the cerebellum and it has. Some cells of the reticular formation of the medulla send their fibres via the restiform body to the cerebellum. Others in the reticular formation of the pons send their fibres to the cerebellum via the brachium pontis. Fibres leave the cerebellum via both the restiform body and the brachium conjunctivum to reach the reticular formation of the medulla, the pons, and the midbrain.

III. LOCATION

The reticular substance of the spinal cord is atypical in that most of it lacks a reticular structure. The cells that serve as correlative neurons (and for this reason are to be considered as part of the reticular substance) make up the grey commissure and the adjacent lateral grey matter that lies between the dorsal and the ventral horns. Where this grey matter contacts the lateral funiculus, a small portion is broken up by fibres and acquires the characteristic reticular structure.

In the medulla the reticular substance is enclosed by the inferior olive ventrally, the medial lemniscus medially, and the motor and sensory nuclei dorsally. Laterally, its borders are the spinocerebellar tracts and the lateral spinothalamic tract. The nucleus ambiguus is embedded in it.

In the pons the reticular substance reaches the midline; dorsally it is limited by the grey matter in the floor of the fourth ventricle; ventrally it contacts the medial and the lateral lemnisci and the superior olive; laterally the lateral lemniscus and the lateral spinothalamic tract border it. The motor nuclei of nerves 5 and 7 are embedded in it.

In the midbrain the reticular substance is enclosed by the medial longitudinal fasciculus and the tectum dorsally, the red nucleus and the medial lemniscus ventrally, the median plane medially, and the auditory pathway and the lateral spinothalamic tract laterally. The reticulothalamic tract, the ascending pathway of the reticular formation, forms a fairly compact bundle in the reticular formation close to the midline and dorsal to the red nucleus.

IV. THE ROLE OF THE RETICULAR SUBSTANCE IN SPASTICITY DUE TO AN UPPER MOTOR NEURON LESION

In concluding this general description of the reticular substance it will be of interest to outline an explanation that has been put forward to explain the spasticity that is associated with an interruption of the pyramidal pathway in the internal capsule. It is suggested that the state of maintained contraction in the muscles which are paralyzed by such a lesion is caused by a loss of the inhibiting influence of the reticular formation of the medulla. Normally, it is assumed, fibres from the cerebral hemisphere exert this inhibition through the reticular formation of the medulla. Such fibres travelling with the corticospinal fibres have been demonstrated anatomically. Thus when the pyramidal pathway is cut, the ventral horn cells are not only deprived of the impulses conveyed to them by the pathway for voluntary movements but are deprived of the modulating influence exerted via the reticulospinal pathways from the medulla. Hence, the impulses reaching the ventral horn cells via the reflex pathways which are still intact and are continually being activated by impulses from the sense organs (for example, in muscles and joints) are not blocked. The result is a maintained contraction in the muscles that are paralyzed as far as voluntary movements are concerned.

9. The Cerebellum

I. SIGNIFICANCE OF THE CEREBELLUM

THE CEREBELLUM is an appendage of the central nervous system in both a structural and a functional sense. Structurally it is attached by a stalk to the back of the hindbrain (figure 71) and functionally it feeds impulses into descending pathways that bypass it in the brain stem. Hence, removal of the cerebellum does not impair sensation or result in paralysis. Still, it is an important accessory structure because without it motor responses are awkward and jerky.

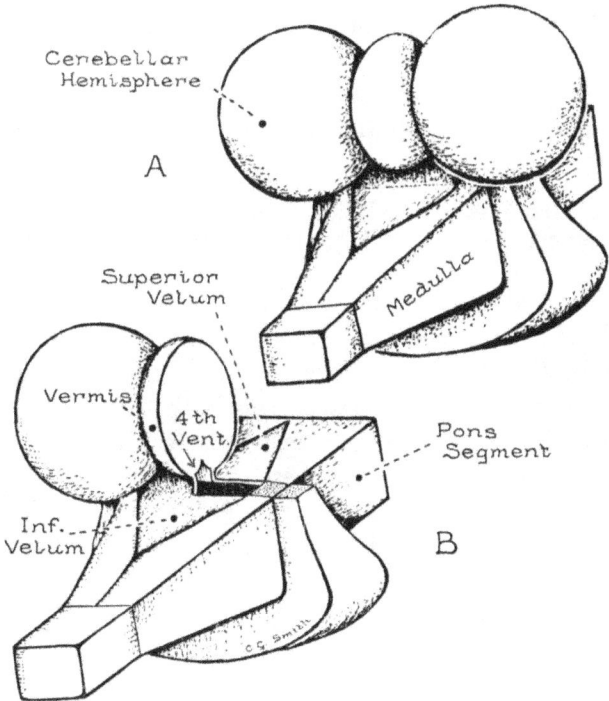

FIGURE 71. A: The dorsolateral aspect of a model of the hindbrain and cerebellum. B: The same as A with the right half of the cerebellum removed.

Its role in motor responses, that is, in reflex or voluntary movements is two-fold: (1) to supply the extra impulses that are required by a muscle to ensure a steady waxing and waning of its contraction and (2) to regulate the flow of impulses to muscles involved in a movement to ensure their working co-operatively.

II. THE FORM AND POSITION OF THE CEREBELLUM

The cerebellum is a dumbbell-shaped mass set across the back of the pons (figure 71). It has a median constricted part called the vermis and right and left larger portions called cerebellar hemispheres. Because it develops as a thickening of the roof of the fourth ventricle, it necessarily forms the part of the roof between the superior velum and the inferior velum. The fourth ventricle has a very short, tent-like dorsal recess in the cerebellum. At the lateral ends of the transverse line of attachment of the cerebellum to the hindbrain, that is, at the lateral edges of the fourth ventricle, we find the stout pedicles of nerve fibres that connect the cerebellum and the brain stem. This is illustrated in diagram 33.

III. THE GREY AND WHITE MATTER AND THE COURSE OF NERVE IMPULSES WITHIN THE CEREBELLUM

The white matter forms the core of the cerebellum and also the right and the left pedicles that connect the cerebellum with the brain stem (figure 72). Most of the grey matter of the cerebellum is in a thin layer on the external surface. This bark-like covering is called the cortex. The rest of the grey matter is in a cluster of centrally located nuclei. There are four of these in the core of each half of the cerebellum, one in the vermis and three in the cerebellar hemisphere. They are close to the fourth ventricle; the medial and lateral ones bulge into it.

The cortex receives most of the fibres that enter the cerebellum; only a relatively insignificant number, which will not be considered here, go to the central nuclei. The central nuclei on the other hand give rise to almost all the fibres that leave the cerebellum; again, only an insignificant number go directly from the cortex to the brain stem. In this circuit through the cerebellum (figure 72), it is the

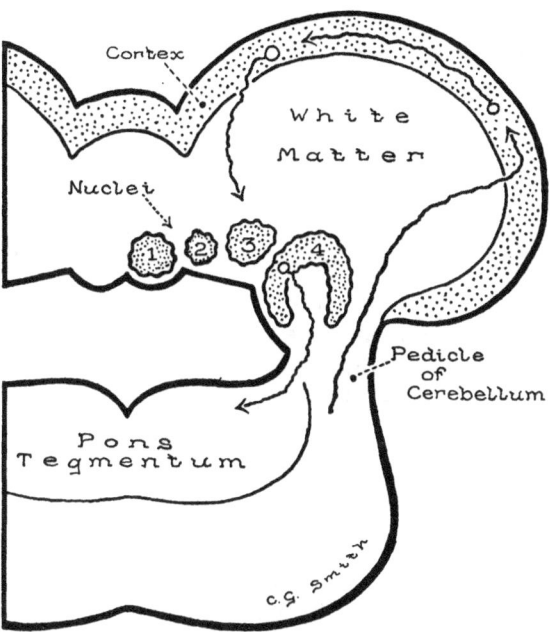

FIGURE 72. A cross-section to show the grey and white matter of the cerebellum and the course of an impulse entering the cerebellum.

cortex that contributes the extra nerve impulses that are required by muscles in the performance of precise movements; thus its volume in man is large. This large mass of cortex has to be accommodated by increasing the surface of the cerebellum, because, as will be explained later, the cortex is organized to have a fixed thickness. Hence, the surface of the cerebellum has a great many furrows or sulci. These furrows, unlike those of the cerebral hemisphere, all extend from side to side. A few of them are very deep and the lamella-like portions between two of these deep furrows are the cerebellar lobules (figure 75). Each lobule has a thick, free, superficial border and a very thin, deep, attached border. This is seen in a midsagittal section (figure 74). The central part of the cerebellum to which the lamellae are attached will be referred to as the core of the cerebellum. The surface of each lobule is, in turn, furrowed by grooves that also extend mediolaterally. These are numerous and of varying depth. They cut into the substance of the lamella in such a way that sagittal sections of it resemble a tree with a trunk, primary and secondary branches, and leaves. The leaves of this "tree," called folia, are the sections of the finest of the mediolateral ridges. These are about 2 mm. wide and cover the whole surface of the cerebellum.

IV. THE SURFACES OF THE CEREBELLUM

At an early stage of development a sagittal section of the cerebellum is circular; later its outline conforms to the cavity it fills (as shown in figure 73). For descriptive purposes the surface of the cerebellum is divided into superior and inferior portions by a horizontal plane through its stalk. As the cerebellum enlarges, the posterior part of its superior surface is pressed flat against the tentorium cerebelli and at the same time its anterior part is pressed against, and is moulded around, the brain stem. The middle of the transverse ridge where these two parts of the superior surface meet is known as the culmen monticuli (ridge of a little mountain) (figure 73A). The inferior surface of the cerebellum rests on the floor of the cup-like posterior cranial fossa and retains its embryonic, rounded contour.

A. THE PART OF THE SUPERIOR SURFACE PRESSED AGAINST THE BRAIN STEM

This surface embraces the brain stem at the junction of the pons and midbrain (figure 73A). It bears the impressions of the inferior colliculi, the brachia conjunctiva, and the crura cerebri. At the midline, between the brachia conjunctiva, it rests on the superior velum.

B. THE PART OF THE SUPERIOR SURFACE PRESSED AGAINST THE TENTORIUM CEREBELLI

This part of the superior surface has the outline and contours of the dorsal surface of a butterfly with drooping, outstretched wings (figure 73B). There is a median ridge corresponding to the body of the butterfly and from it extend the flat, sloping surfaces of each hemisphere that correspond to its wings. There is no line here to mark the junction of vermis and hemisphere. The anterior border of this surface has a large, deep notch for the brain stem. To either side of this median notch the anterior border is in contact with the petrous bone and is therefore straight and inclined backwards. The posterior border of this surface also has a notch in the midline, a small one,

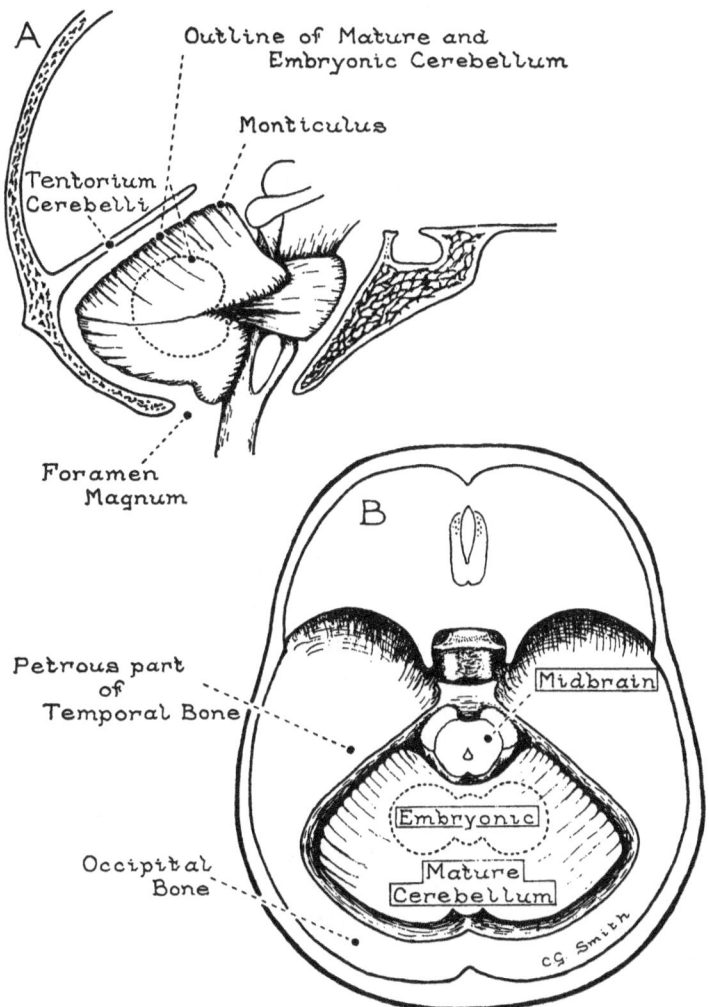

FIGURE 73. A: The lateral aspect of the cerebellum and brain stem in a midsagittal section of the skull. B: The superior aspect of the cerebellum and the caudal part of the brain stem fitted into the base of the skull.

for the internal occipital crest. To the right and left of this notch the posterior border is convex, moulded by the occipital bone.

C. THE INFERIOR SURFACE

This surface has the same outline as the superior surface but its contours are quite different (figure 76). The vermis lies in a deep notch, or *vallecula* (valley) between the hemispheres that are round, not flat. This surface fits the mould formed by the median internal occipital crest and the cup-like occipital bone to either side of it. Where the cerebellum is unsupported, that is, over the foramen magnum, it bulges downwards into the vertebral canal (figure 73A). This herniation may be large or small and, as we shall see later, involves the tonsillar lobule and also part of the biventral lobule. The lateral border of the convex inferior surface and the lateral border of the flat part of the superior surface are separated by a cleft that has the brachium pontis in its floor (figure 73A). The width of this cleft increases toward the brain stem.

V. THE LOBULES OF THE CEREBELLUM

As pointed out above, deep transverse fissures divide the cerebellum into a rostrocaudal series of lobules. These lamella-like lobules extend from one side of the cerebellum to the other but some become very slender at the junction of the vermis and the cerebellar hemisphere. Figures 74A and 74B picture the lobules in a diagram of the surface of the right half of the cerebellum. In this diagram the worm-like vermis is represented as uncoiled to make all its lobules visible in a dorsal view. The lateral portions of the lobules, which are located in the cerebellar hemisphere, are similarly pictured as lying on a flat surface. They are labelled using the nomenclature of human anatomy and also according to the nomenclature of comparative anatomy.

As shown in figure 74B the rostral lobule, the **lingula**, is restricted to the vermis. It is a tongue-shaped lobule that appears to have come into being by an extension of cortex onto the superior velum (figure 74A). Actually it is a lamella that is formed by the development of a fissure on the rostral part of the superior surface of the embryonic cerebellum. The lingula can be seen to advantage in the dissection illustrated in figure 75B.

The next lobule in the rostrocaudal series is the **central lobule**. The vermal part of this lamella is thick, its lateral parts are wing-like, short, and taper to a thin edge. It is located on the part of the superior surface of the cerebellum that is applied to the brain stem and is attached to the core of the cerebellum just behind the lingula, in the vermis, and at the anterior border of the core of white matter in each hemisphere (figure 75B).

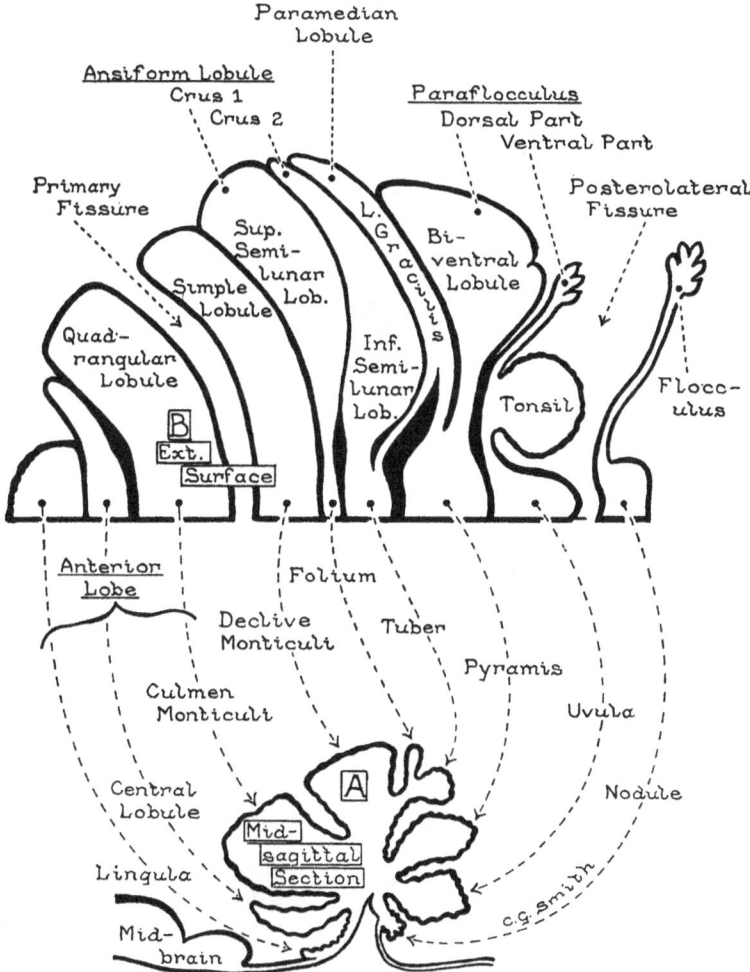

FIGURE 74. The lobules of the cerebellum. A: A midsagittal section. B: A schematic drawing of the surface of the right half of the cerebellum.

The next lamella is a broad band with a crest that forms the border of the notch for the midbrain in a dorsal view of the cerebellum (figure 75A). The vermal portion of this lamella forms the high point of the midsagittal section of the cerebellum and is called the **culmen monticuli**. The lateral portion of this lamella is called the **quadrangular lobule**. The fissure at the posterior border of this lamella, that is, behind the culmen and the quadrangular lobule, is the **primary**

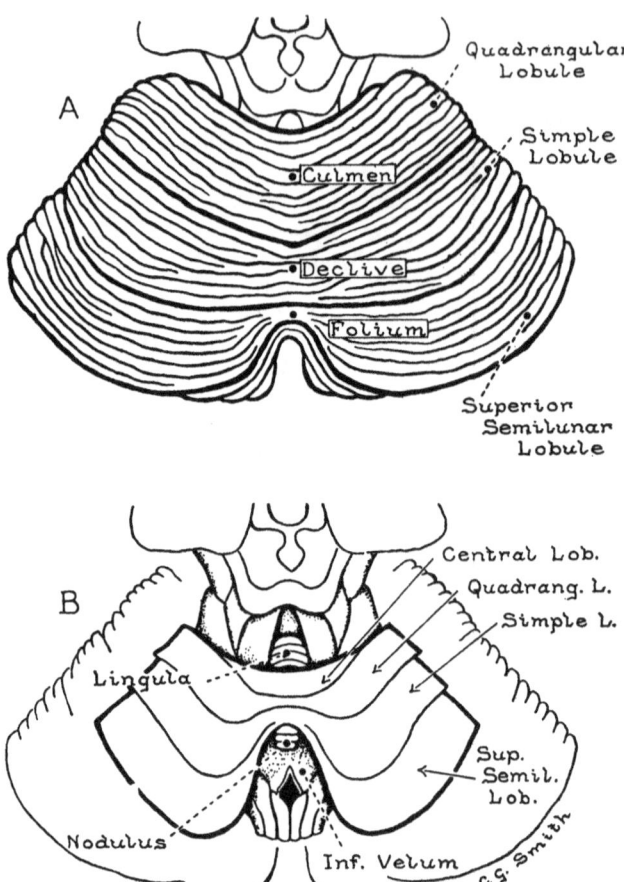

FIGURE 75. A: Superior aspect of the cerebellum. B: The same as A with most of the lobules detached to show the white core and the lingula and the nodulus.

fissure. This was named before it was known that another, the posterolateral fissure (to be described later), preceded it in the course of development. Behind the primary fissure is a broad lamella with a uniform width. The vermal portion is called the **declive monticuli** (slope of a little mountain); the hemispheric portion is called the **simple lobule**. Note that in the terminology of comparative anatomy, the simple lobule includes the vermal and hemispheric portions. The declive and the culmen monticuli form almost the whole of the vermal portion of the tentorial surface of the cerebellum.

The declive and the next three cerebellar lamellae caudal to it share one stout branch of the arbor vitae in the vermis (figure 74A). Only in the cerebellar hemisphere do these four lamellae acquire independent attachments to the core of the white matter and qualify as independent lobules. The vermal portion of the lobule immediately behind the declive is the **folium**. This thin lamella grows wider as it is followed into the hemisphere where it is called the **superior semilunar lobule**. It lies along the posterior border of the superior surface. This, the simple lobule, and the quadrangular lobule each form about one-third of the tentorial surface of the hemisphere (figure 75A).

Next to the folium, at the posterior border of the inferior surface of the vermis (figure 76A) is the tubercle-like composite mass called the **tuber vermis**. It includes the attenuated midline portions of two

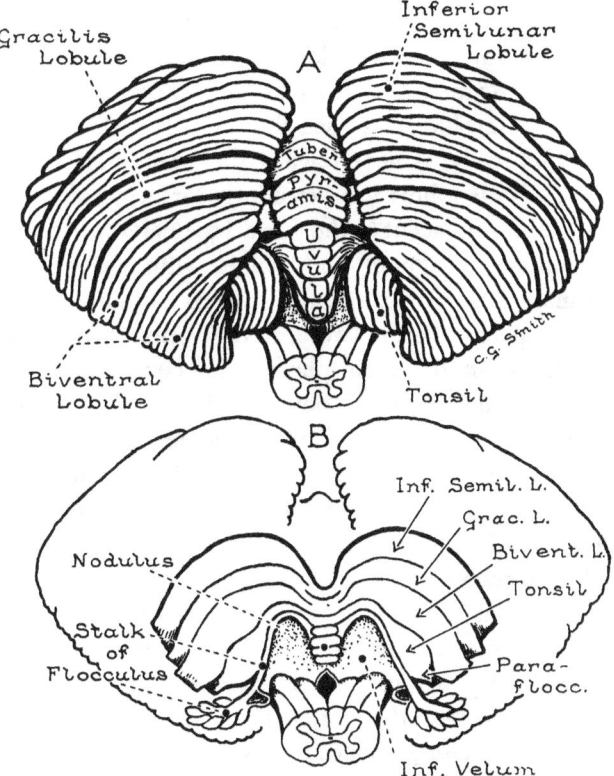

FIGURE 76. A: The inferior aspect of the cerebellum. B: The same as A with most of the lobules detached to show the white core and the flocculonodular lobe.

lobules that are large and distinct in the hemisphere. They are, in order, the **inferior semilunar lobule**, thick at its medial end and tapered laterally and the slender **gracilis lobule**, which has a uniform thickness throughout its length. The fissure between the superior and inferior semilunar lobules is the **horizontal fissure** (not labelled in figure 74). It is close to the posterior border of the inferior surface and corresponds to the fissure which lies between crus 1 and crus 2 of the ansiform lobule in lower animals. At its lateral end it is continuous with the wide, horizontal fissure which has the brachium pontis in its floor.

The **pyramis** (pyramid), the next lobule of the vermis, forms an eminence on the inferior surface. It is the median part of a lamella that is considerably attenuated at the junction of the vermis and hemisphere. Laterally it enlarges to form the massive **biventral lobule** (two bellies) but it has, in addition, a slender connection with the gracilis lobule. The biventral lobule (lamella) becomes so thick toward its lateral extremity that its caudal border, that is, the border nearer the medulla oblongata (see figures 74B and 76A), instead of extending transversely, is bent toward the brain stem. This explains why the folia of this lobule are extending dorsoventrally. The part of the biventral lobule adjacent to the tonsil (see below) lies unsupported above the foramen magnum and bulges with the tonsil into the vertebral canal. The next lamella in the rostrocaudal series (figure 74B) has three parts, viz. the **uvula**, in the vermis, and the **tonsil** plus the **paraflocculus**, in the hemisphere. The connection between the uvula and the tonsil and again between the tonsil and the paraflocculus is so slender that it is difficult to demonstrate in a dissection. The three parts are, however, continuous and have an unbroken line of attachment to the core of the cerebellum as shown in figure 76B.

In figure 76A the resemblance of the uvula to the uvula of the palate and the resemblance of the tonsil to the palatine tonsil is striking. Notice that the folia of the tonsil are bent into the sagittal plane as are the ones adjacent to it in the biventral lobule. The tonsil is pressed anteriorly against the inferior velum (figure 77B) and inferiorly it bulges into the vertebral canal along the side of the medulla oblongata. The paraflocculus is a tuft of three or four folia at the end of a slender band that is prolonged laterally from the tonsil. These folia are located just behind the folia of the flocculus

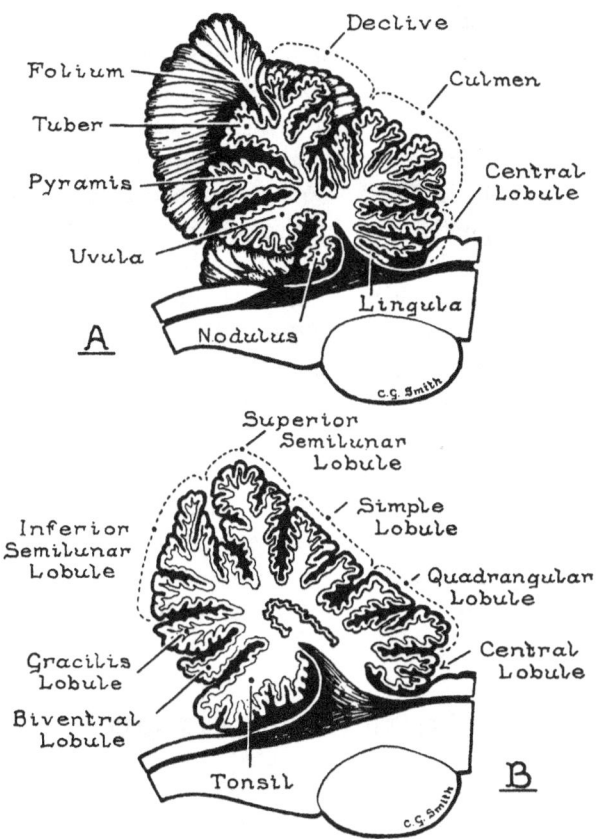

FIGURE 77. A: A sagittal section of the vermis. B: A sagittal section of the cerebellar hemisphere, 1 cm. from the median plane.

and can not be distinguished from them except by dissection. The area of attachment is shown in figure 76B.

The last of the rostrocaudal series of transverse lamellae is called the **flocculonodular lobe**. As its name implies it has two parts, the **nodule**, in the vermis, and the **flocculus**, in the hemisphere. This very small structure has the status of a major subdivision of the cerebellum. It is, indeed, the original cerebellum. The **posterolateral fissure** which separates it from the rest of the cerebellum is the first fissure to appear in both embryological and phylogenetic development. The nodule in the vermis is a small mass of folia about the size of an orange seed. The flocculus is a tuft of four or five bud-like folia at the lateral end of the lamella. Between the nodule and flocculus the lamella is reduced to a delicate band of fibres called the stalk of the flocculus. This band lies along the line of attachment of the inferior velum and, as illustrated in figure 76B, it has a sharp bend just lateral to the nodule as it courses laterally on the surface of the core of the cerebellum. The flocculus is applied to the caudal border of the brachium pontis. The myelinated fibres of the stalk of the flocculus commonly take a short-cut from the nodule to the flocculus travelling in the inferior velum and thus convert it into a medullary velum (a velum containing nerve fibres).

In the above description of the lamellae of the cerebellum occasional reference has been made to the names given to these lamellae in other mammals as shown in figure 74B. Some of these correlations between man and other animals must be considered as tentative. The form and arrangement of the subdivisions of the human cerebellum are so different from the pattern found in other mammals that additional detailed studies of development and of fibre connections will be required to relate with certainty the lamellae of man and animals.

VI. THE CORE OF THE CEREBELLUM

The core of the cerebellum is the central mass of white matter to which the lobules of the cerebellum are attached. It contains the central nuclei. In order to expose the core of the cerebellum as in figures 75B and 76B each lobule has to be detached with a knife. When this is done the central core is seen to have a dumbbell-shape as does the cerebellum as a whole. It consists of right and left ovoid masses united across the midline by a narrow, thin band of white matter. It has two surfaces, a superior or dorsal and an inferior or ventral. Sagittal sections of the vermis and of the cerebellar hemisphere (figure 77) show that the whole of the superior surface and the posterior part of the inferior surface provide attachment for lobules. The anterior part of its inferior surface is in the roof of the fourth ventricle. The major fibre bundles of the core and the nuclei embedded in the core can be exposed by dissection as shown in figures 78 and 79.

In figure 78B the lobules have been detached and with them

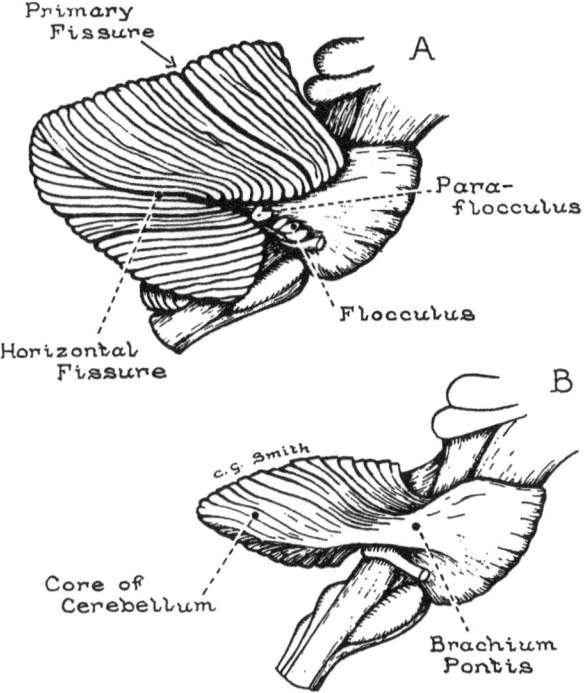

FIGURE 78. Steps in the dissection of the cerebellum. A: The lateral aspect of the undissected cerebellum. B: The same as A with the lobules detached to show the contribution of the brachium pontis to the core.

enough of the superficial fibres have been teased away to show the course of the fibres of the brachium pontis. The latter form the lateral margin of the core of white matter before they course medially to reach all lobules except the flocculonodular lobe. In figure 79A the brachium pontis has been cut across close to the brain stem and its fibres have been removed. This exposes the restiform body which makes a sharp turn dorsally just rostral to the cochlear nuclei and enters the stalk of the cerebellum between the two other large bundles in its stalk, that is, the brachium pontis lateral to it and the brachium conjunctivum medial to it. As it turns dorsally its fibres also fan out to reach all parts of the cortex of the cerebellum. This has been shown to be true of the olivocerebellar fibres in the restiform body; other fibres in this bundle, such as the spinocerebellar fibres, reach a part, not all, of the cerebellar cortex.

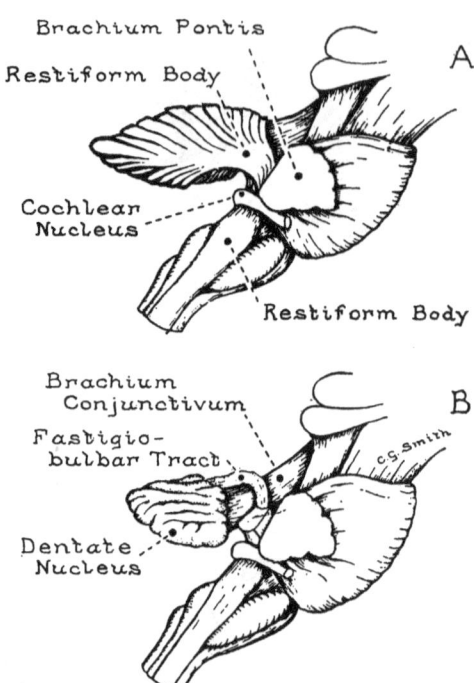

Figure 79. Dissections of the cerebellum. A: The fibres of the brachium pontis have been peeled off the core revealing the restiform body. B: The fibres of the restiform body have been removed exposing the dentate nucleus.

The next step in the dissection is to strip away the fibres of the restiform body. If these are removed a few at a time beginning laterally, a bundle of fibres called the fastigiobulbar tract can be demonstrated. The fibres of this bundle do not fan out like the other fibres of the restiform body, but instead course medially as a compact bundle along the anterior border of the dentate nucleus to reach the fastigial nucleus. The fastigial nucleus contains the cell bodies of these fibres and is located in the vermis medial to the other nuclei of the core.

The removal of the restiform body exposes the capsule of fibres that forms the immediate covering of the central nuclei. The capsule is formed by fibres coming to the nuclei from the cortex. With the removal of these encapsulating afferent fibres the central nuclei are exposed (figure 79B). Let us examine these nuclei beginning with the largest and most lateral one, the dentate.

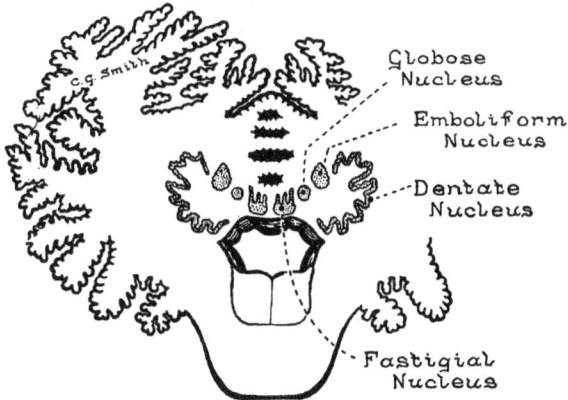

FIGURE 80. A cross-section of the cerebellum and pons to show the cerebellar nuclei.

The dentate nucleus. This is a thin lamella of grey matter that has the form of an inverted, wrinkled sac with its large opening directed medially and rostrally (figures 77B and 80). The lamella is about one-third the thickness of the cortex, about ⅓ mm. It resembles the cortex in that its surface has many folds but unlike the cortex its cells are not in layers and its grooves and ridges are in the sagittal plane (figure 79B). The efferent fibres of the nucleus form its core and leave its rostral, open end as a compact, ribbon-like bundle, the brachium conjunctivum. The medial surface of this bundle is covered by the ependyma of the fourth ventricle (figure 77B). The postero-lateral rim of the cup-like dentate nucleus is close enough to the ventricular cavity to form an elevation on its wall. The nucleus is about 2 cm. long, about 1 cm. wide, and a little less than 1 cm. deep.

Lying along the dorsomedial border of the dentate nucleus is the elongated, stopper-shaped **emboliform nucleus** (*embolus*, a stopper) (figure 80). It is so called because it resembles a stopper for the sac-like dentate nucleus. This relationship to the dentate can be appreciated by referring to the drawings of figures 80 and 81. The nucleus has a bulbous, anterior end that lies on the medial side of the brachium conjunctivum as it emerges from the dentate nucleus. Caudally it tapers to a point. It is about 2 cm. long and about ½ cm. in diameter near its anterior end. The nucleus is irregular in outline and it is connected with the dentate nucleus at one or two points along its length. Its cells like those of the dentate nucleus are

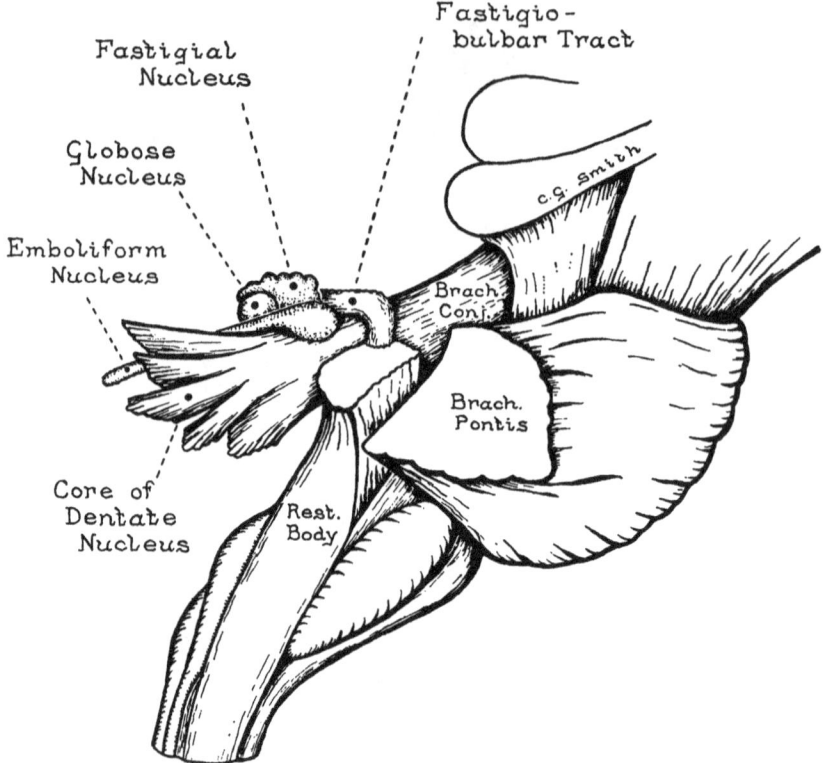

FIGURE 81. The last step in the dissection of the cerebellum. The dentate nucleus has been removed from the specimen shown in figure 79B.

large and multipolar; the axons of these cells join the brachium conjunctivum.

The next of the mediolateral series of central nuclei is the **globose nucleus** (figures 80 and 81). This is not a single globular mass but a series of two or more small, spherical masses of grey matter located along the medial side of the emboliform nucleus. They vary greatly in size and number but they may attain a diameter of ½ cm. The cells are multipolar, some large, some small, and their axons join the brachium conjunctivum and the fastigiobulbar tract.

The most medial of the four central nuclei is the **fastigial nucleus** or nucleus tecti, that is, the roof nucleus (figures 80 and 81). This nucleus is in the vermis close to the median plane and as its name indicates it is adjacent to the roof of the fourth ventricle. It is a dorsoventrally flattened ovoid mass with root-like processes projecting dorsally and posteriorly. It is about 1 cm. long and extends from the lingula to a point close to the attachment of the pyramis. It is about ½ cm. wide and ½ cm. thick. Some of its cells resemble those of the other central nuclei but it also contains cells that resemble the vestibular nuclei. This, plus the observation that there is a bridge of scattered cells in the wall of the ventricle between the fastigial and the superior vestibular nucleus, has been taken to indicate that the fastigial nucleus is in part a dorsal extension of the vestibular nuclei. Most of the cells of the fastigial nucleus send their axons to the vestibular and reticular nuclei of the hindbrain. These fibres form the compact fastigiobulbar tract which arches around the dorsolateral surface of the brachium conjunctivum to reach the brain stem along the medial side of the restiform body.

With the exposure of the central nuclei we have completed the dissection of the cerebellum. We have proceeded step by step from its external surface, through its core of white matter and its stalk, to the fourth ventricle and have identified all of its parts. The next step is to assimilate the data made available in embryological, pathological, and experimental studies concerning the origin and termination of the fibres that make up its pathways.

VII. THE PATHWAYS THROUGH THE CEREBELLUM

In tracing the pathways through the cerebellum we will consider, in turn, the source and distribution of the afferent fibres of the cortex, the course of the impulses through the cortex, the efferent fibres of the cortex to the central nuclei, and lastly the efferent fibres of the central nuclei.

A. THE AFFERENT FIBRES OF THE CEREBELLAR CORTEX

1. AFFERENTS FROM SENSE ORGANS OF THE INTERNAL EAR

These are either fibres of the vestibular nerve or fibres from the vestibular nuclei. They enter the cerebellum on the medial side of the restiform body. The primary vestibular nerve fibres reach the nodule and the flocculus. The secondary vestibular fibres (from the vestibular nuclei) end in the cortex of the lingula, the uvula, the nodule, and the flocculus.

2. AFFERENTS FROM SENSE ORGANS OF THE SPINAL NERVES

The pathways that carry impulses into the cerebellum from the trunk and the limbs are (a) the dorsal and (b) the ventral spinocerebellar tracts, (c) the dorsal external arcuate fibres, (d) a reticulocerebellar tract from the lateral part of the reticular formation of the medulla, and (e) a fibre bundle from the accessory olivary nucleus, a medial portion of the inferior olivary nucleus. All five fibre bundles enter the cerebellum in the restiform body. Some of the fibres of the ventral spinocerebellar tract may not join the restiform body until after it has entered the cerebellum. These aberrant fibres appear to overshoot the stalk of the cerebellum as they ascend and then turn back on the lateral side of the brachium conjunctivum to join the restiform body.

All the fibres of these pathways from the trunk and limbs end in the cortex of the vermis except the reticulocerebellar which also project to the paramedian lobule. The pathways from the right side of the body reach the right side of the cerebellum and those from the left side of the body go to the left side of the cerebellum.

3. AFFERENTS FROM THE CEREBRAL CORTEX

Although the pathways from the cerebral cortex to the cerebellum begin in almost all parts of the cerebral cortex most of them begin in the cortex around the central sulcus. These pathways have a relay station either in the pontine nuclei or in the inferior olive. The corticopontine pathway is in the crus cerebri and the pontocerebellar fibres from the cells in the pontine nuclei cross the midline to enter the cerebellum in the brachium pontis. These fibres reach all parts of the cerebellar cortex except the flocculonodular lobe.

The fibres of the pathway from the cerebral cortex that relay in the inferior olive descend in the reticular formation of the brain stem. Fibres from one hemisphere reach both the right and the left olivary nuclei. The fibres from the cells in each inferior olive stream out of the hilum of the inferior olive and cross the midline to enter the cerebellum in the restiform body. The distribution of these fibres differs from that of the pontocerebellar fibres since they reach only certain parts of the cerebellar hemisphere. Note also that each cerebral hemisphere activates both the right and the left inferior olives and therefore both the right and the left halves of the cerebellum.

4. AFFERENTS FROM THE CAUDATE NUCLEUS, THE GLOBUS PALLIDUS, THE RED NUCLEUS, AND THE GREY MATTER AROUND THE AQUEDUCT

Cells in each of these structures send fibres through the reticular formation to the inferior olive. The caudate and the globus pallidus (both parts of the cerebral hemisphere) send fibres to both the right and left olives as the cerebral cortex does. The fibres of the cells in the inferior olive leave the hilum of the nucleus, cross the midline, and enter the cerebellum in the restiform body. Each of the pathways— from the caudate nucleus, the globus pallidus, the red nucleus, and the grey matter around the aqueduct—terminate in different parts of the cerebellar cortex.

5. AFFERENTS FROM THE SPECIAL SENSE ORGANS OF THE EYE AND THE COCHLEA

Oscillographic recording has shown that impulses from these sense organs reach the tuber and the folium. It is believed they come by way of the tectum of the midbrain and are conveyed by the tectocerebellar tract, recognized in the brains of lower animals and in the embryonic human brain. The tract enters the cerebellum by way of the superior (anterior) medullary velum.

In summary, the afferent fibres of the cerebellar cortex fall into two groups: one group conveys impulses from sense organs and the other conveys impulses from the cerebral hemisphere and the midbrain. The pathways from sense organs end in the vermis. The pathways from the cerebral hemisphere and the midbrain end in either the vermis or the hemisphere of the cerebellum or in both, each pathway having its own region of termination.

B. THE PATH OF IMPULSES PASSING THROUGH THE CEREBELLAR CORTEX

The cerebellar cortex is a superficial layer of grey matter with the cells arranged in layers. Its structure is relatively simple. It does not vary from lobule to lobule and what is more remarkable it has the same structure in all animals. The cortex of the fish looks the same as that of man. It would appear therefore that it does the same work in all animals including man, that is, feed impulses into the motor pathways to regulate motor responses.

As shown in figure 82 the cortex has three layers. The middle layer, one cell thick, contains the large efferent Purkinje cells. These are cone-shaped with a thick, apical dendrite directed toward the surface. This dendrite branches repeatedly like a tree but all its branches are in one plane, the sagittal plane. This fan-shaped tuft of dendrites reaches as far as the external border of the cortex. The thickness of the superficial layer of the cortex is therefore equal to the length of the dendrite of the Purkinje cell, that is, about ½ mm. In a sense the

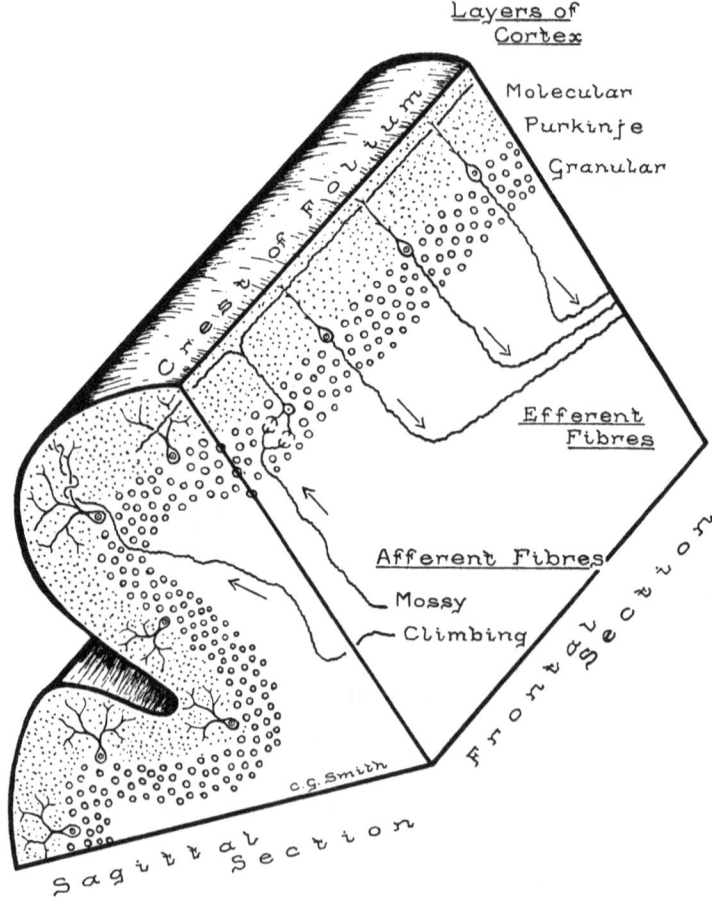

FIGURE 82. A rectangular block from the surface of the cerebellum to show the form of the Purkinje cells in a frontal and in a sagittal section.

thickness of the outer layer of the cortex is determined by the length of the dendrites of the cells in the second layer.

The deepest layer of the cortex is the granule cell layer which contains the afferent cells of the cortex. They resemble granules because the cytoplasm is scanty and the nucleus alone is seen in Nissl preparations. These cells have a tuft of delicate dendrites which receive the moss-like endings of the afferent fibres. Each afferent fibre synapses with several granule cells. The unmyelinated axon of each granule passes into the superficial layer and there it divides into two branches that course parallel to the surface in opposite directions. One branch courses toward the median sagittal plane and the other branch courses away from it. Each of the branches passes through the fan-shaped tuft of dendrites of many Purkinje cells, synapsing with each one in passing. The arrangement achieves two ends: (1) it makes it possible for the axon of one granule cell to activate a large number of Purkinje cells, and (2) it makes for economy of cells in that one group of Purkinje cells can be utilized in turn by many different granule cells. The orderly folding of the cortex, with the sulci all transverse, is probably an adaptation to permit the axons of the granule cells to course in straight lines.

In concluding this sketchy outline of the organization of the cerebellar cortex it is necessary to point out that there is another more direct pathway for impulses passing through the cortex. The afferent fibres of this second pathway are called climbing fibres. They bypass the granule cells and synapse with the Purkinje cell, climbing its dendrite like a vine. It is suggested that these fibres may be branches of the axons of Purkinje cells or branches of the axons of cells in the central nuclei and that they are a part of a feed-back mechanism.

C. THE EFFERENT FIBRES OF THE CORTEX

This part of the pathway through the cerebellum may be described as the projection of the cortex onto the central nuclei. This projection is an orderly one. In a general way the axons of the Purkinje cells in any sagittal plane project to those cells of the central nuclei that are located in the same sagittal plane. Thus the cortex of the vermis projects exclusively to the fastigial nucleus; the cortex of the narrow sagittal band adjacent to the vermis projects to the globose and emboliform nuclei (these are not separate nuclei in the brains of experimental animals) and the large lateral portion of the cerebellar hemisphere projects to the dentate nucleus. (Note—the flocculus, a part of the hemisphere, does not project to the dentate nucleus. Its efferent fibres go directly to the vestibular nuclei). In each plane the anterior and posterior fibres of the cortex reach the anterior and posterior portions of the central nuclei respectively.

D. THE EFFERENT FIBRES OF THE CENTRAL NUCLEI

The axons of the dentate, the emboliform, and the globose nuclei leave the cerebellum in one bundle, the brachium conjunctivum. The fibres cross the midline in the decussation of the brachia conjunctiva and end partly in the red nucleus and partly in the lateral nucleus of the thalamus. From the red nucleus impulses are carried to the anterior horn cells by the rubroreticular and the reticulospinal fibres. From the thalamus, impulses are carried to the motor cortex in the precentral gyrus by fibres of the thalamic radiation. The latter impulses appear to regulate the activity of the cortex because destruction of this pathway to the motor cortex is followed by spontaneous involuntary movements.

The axons of the cells of one fastigial nucleus are joined by some fibres from the fastigial nucleus on the opposite side and together they pass laterally on both the dorsal and ventral aspects of the brachium conjunctivum to reach the tegmentum of the pons. They reach the vestibular and reticular nuclei of the brain stem and constitute the fastigiobulbar tract. The fibres that pass dorsal to the brachium conjunctivum are the fibres of the compact bundle described in the dissection of the core of the cerebellum and illustrated in figures 79B and 81. It arches laterally just in front of the dentate nucleus and forms the medial part of the restiform body.

In concluding this summary of the composition of the pathway through the cerebellum it is important that we should appreciate that there are significant exceptions to the general scheme as outlined above. (1) Some of the axons of the Purkinje cells of the flocculonodular lobe pass directly to the vestibular nuclei and the reticular nuclei of the brain stem. (2) Afferent fibres of the cerebellum end not only in the cortex but in the central nuclei as well. Experimental evidence is available to show that this pathway can function after the cortex is destroyed.

VIII. THE LOBES OF THE CEREBELLUM

Thus far in our description of the cerebellum we have carefully avoided committing ourselves concerning the division of the cerebellum into lobes. The terms anterior lobe and flocculonodular lobe have been used in describing the cerebellum but only as proper names for the portions of the cerebellum located in front of the primary fissure and behind the posterolateral fissure, respectively. It has been our purpose to describe the form, position, and connections of the different parts of the cerebellum designating each by name rather than as a part of a lobe. In doing this it was hoped to avoid the confusion arising out of the lack of uniformity in subdividing the cerebellum. Having completed the description we are now prepared to consider the different methods of subdividing the cerebellum that have been proposed.

One attractive method of subdividing the cerebellum was first advanced by Ingvar in 1918. He suggested dividing the cerebellum into parts on the basis of its afferent fibres. According to this method, the cerebellum has three parts receiving impulses: one from vestibular sense organs, another from sense organs of spinal nerves, and a third from the cerebral cortex. Since, phylogenetically, the cerebellum develops as an outgrowth of the vestibular nuclei, the vestibular portion is the oldest and is called the **archicerebellum**; this is the flocculonodular lobe. The second portion of the cerebellum to develop came into being to receive the impulses from the spinal cord. Included in this are the vermis portion of the anterior lobe, the vermis portion of the simple lobule, the pyramis, and the uvula. This then is the second oldest part of the cerebellum and has been called the **paleocerebellum**. The third and the newest part of the cerebellum, the remainder, is the part that receives impulses from the cerebral cortex; this is the **neocerebellum**.

In spite of its attractiveness the method of dividing the cerebellum into parts on the basis of its afferents has been criticized because the areas receiving these afferents overlap. The fibres from the spinal cord and from the vestibular nuclei overlap in the uvula and the lingula, and the fibres carrying impulses from the cerebral cortex reach all parts of the cerebellar cortex. This method of dividing the cerebellum also encounters the criticism that it includes the folium and the tuber in the neocerebellum. These are parts of the vermis and therefore are not phylogenetically new. Moreover, they share a characteristic of the rest of the vermis, namely, that they receive impulses from sense organs. The sense organs in this case are the special ones of hearing and seeing (oscillographic studies).

A more acceptable method of dividing the cerebellum into lobes is that of Larsell. He subdivides the cerebellum on the basis of both its development and its fibre connections. According to this method the cerebellum is divisible into two basically different parts, namely, the flocculonodular lobe and the corpus cerebelli. The flocculonodular lobe is set apart because it receives direct vestibular nerve fibres, like the vestibular nuclei. Furthermore the posterolateral fissure, which separates these two parts of the cerebellum, is the first fissure to develop. Larsell divides the corpus cerebelli, in turn, into anterior and posterior lobes using the primary fissure, the first fissure to appear in the developing corpus cerebelli. The anterior lobe receives most of the spinocerebellar fibres and the posterior lobe (of the corpus cerebelli) receives most of the pontocerebellar fibres. Unfortunately the anterior lobe, thus defined, does not conform to the anterior lobe as charted by the physiologist. The physiologists would include the simple lobule in the anterior lobe.

A fundamentally different method of subdividing the cerebellum is based on the projection of the efferent fibres of its cortex. According to this method the cerebellum is divided into a mediolateral series of three sagittal segments. The medial portion of each half of the cerebellum corresponds to the vermis. It sends its efferent fibres to the fastigial nucleus. The medial portion of the hemisphere is the second subdivision of the cerebellum. It sends its fibres to the globose and emboliform nuclei. The third subdivision is the lateral portion of the hemisphere. It projects to the dentate nucleus.

Recent studies of function support such a division. It was found, for example, that lesions in any part of the vermal zone impaired to some extent the ability to maintain equilibrium, while lesions in the paravermal, that is, the intermediate portion, impaired postural adjustments of the limbs on the same side. No specific function of the lateral zone was revealed by stimulating or destroying it, but in man it is known to play an important part in the smooth performance of voluntary movement.

IX. FUNCTION OF THE CEREBELLUM

The cerebellum as stated earlier is an accessory structure of the brain connected to the pathways that descend to the motor nuclei. It does not initiate movement, and therefore removing it does not result in paralysis. In terms of muscular physiology it does two things. It supplies the extra nerve impulses that are required to ensure a steady waxing and waning of a contraction of a given muscle, and it regulates the flow of impulses to muscles sharing in a response. Thus if a part or all of the cerebellum is destroyed, one may expect the muscles to be softer than normal, that is, lacking in tone, and one may expect voluntary movements to be awkward and executed with a tremor. Starting with this basic statement of function the question arises as to whether there is a part of the cerebellum for each part of the body such as the head, the arm, or the leg.

Such a relatively simple concept of localization in the cerebellum led comparative anatomists to make shrewd guesses based on the size of each of the parts of the cerebellum in animals such as the giraffe (long neck), the whale (rudimentary limbs), and man (emancipated upper limb). Such guesses were, however, only partly supported by experimental studies and the observations in clinical cases. Indeed, these experimental studies suggested that the localization was primarily such that there was a part of the cerebellum for each of the three kinds of motor activity, namely, equilibrium, posture, and voluntary movement. Thus, destruction of the nodulus impaired the ability to maintain an upright posture. This is well illustrated in children with a tumor involving the nodulus, and also in monkeys with experimental lesions in the nodulus. The child or the monkey may not be able to stand without swaying but there is no tremor, or weakness of the extremities in bringing food to the mouth when at rest on the back. The vermal part of the anterior lobe, on the other hand, and the pyramis and uvula would appear to be concerned with the flow of impulses to the motor pathways controlling posture, which is basic to all voluntary movement. Thus electrical stimulation of the cortex of the anterior lobe in experimental animals will abolish the excessive contraction of the extensors in decerebrate rigidity. On the other hand, removing the cortex of the vermal part of the anterior lobe in animals will result in abnormal distribution of impulses to muscles, resulting in increased tone in extensors so severe as to produce a marked dorsal retraction of the head and arching of the back. Another abnormality will be hyperactive reflexes such as hyperextension of a limb when it contacts a supporting surface. In spite of such a gross deficit, the animals can stand without swaying and can use the forelimbs without a tremor.

An unexpected feature of the findings cited above is the absence of the anticipated hypotonia. The general function of the cerebellum as stated earlier suggests there should be a weakening of muscular contraction following a destruction of cerebellar cortex. Evidently the anterior lobe, pyramis, uvula, and the flocculonodular lobe are chiefly concerned with the regulation of the flow of impulses to opposing muscle groups, and only secondarily concerned in the supply of extra impulses. However, destruction of the cortex of the cerebellar hemisphere in man is followed by both the expected hypotonia and the tremor during voluntary movement. There is no disturbance of equilibrium in an individual with a lesion limited to the cortex of the cerebellar hemisphere.

To summarize the above functional analysis we may note: (1) the flocculonodular lobe is chiefly concerned with equilibrium. This is nicely correlated with the termination there of the pathways from the internal ear. (2) The vermis of the anterior lobe together with the pyramis and uvula form a unit that is chiefly concerned with postural adjustments. This is correlated with the finding that the pathways from the spinal cord end here. Finally, (3) the cerebellar hemisphere, that is, the lateral parts of the anterior and posterior lobes (Larsell), is chiefly concerned with the execution of voluntary movements. This is correlated with the termination there of the large pathways from the cerebral cortex.

In concluding this outline of the function of the cerebellum it is important to take note of the fact that there are receiving areas in the cerebellar cortex for impulses from tactile, auditory, and visual sense organs. The impulses from the eye and from the cochlea reach areas that are coextensive and located in the folium and tuber. The impulses from each cutaneous area reach two separate areas of cerebellar cortex. Those from the head end in both the simple lobule and the anterior part of the paramedian lobule. Those from the upper limb reach both the quadrangular lobule and the middle part of the paramedian lobule. Lastly, those from the lower limb reach both the central lobule and the posterior part of the paramedian lobule. How the impulses reach the paramedian lobule is not known but probably it is via the lateral reticular nucleus of the medulla oblongata. It is also puzzling to find that anatomical studies of the spinocerebellar projections do not show the localization that electrical recording of impulses demands.

10. The Cerebral Hemisphere II: Development, Form, and Structure

THE TWO CEREBRAL HEMISPHERES are, in a sense, accessory parts of the brain. They are attached to the side of the diencephalon, that is, to the rostral end of the axial portion of the brain which we call the brain stem. They have been added to the central nervous system to serve as dominant association mechanisms. Their function is to process the data picked up by all the sense organs, record that data for future reference, and then initiate an appropriate response in the light of that data and the data of past experience.

I. THE DEVELOPMENT OF THE CEREBRAL HEMISPHERE

A brief outline of the development of the cerebral hemisphere will help us to understand its form, structure, and relationships. As indicated above, the cerebral hemisphere has evolved to serve as a dominant association mechanism. To play this role, it develops, as might be expected, close to the rostral end of the brain. This, the most suitable location for an association mechanism, happens to be just behind the attachment of the olfactory nerve. Hence the new growth of nervous tissue destined to become the cerebral hemisphere begins in the olfactory bulb, a vesicle formed by the outpouching of the lateral wall of the telencephalon (end brain). Figure 83A illustrates an early stage in the development of the hemisphere. In this illustration, the vesicle formed by the evagination of the lateral wall of the telencephalon which originally was entirely olfactory bulb is already divisible into two parts: a larger part, the future cerebral hemisphere, and a smaller part, the olfactory bulb proper. The latter, as development proceeds, extends toward the roof of the nose where the olfactory nerve fibres enter the cranial cavity and its connection

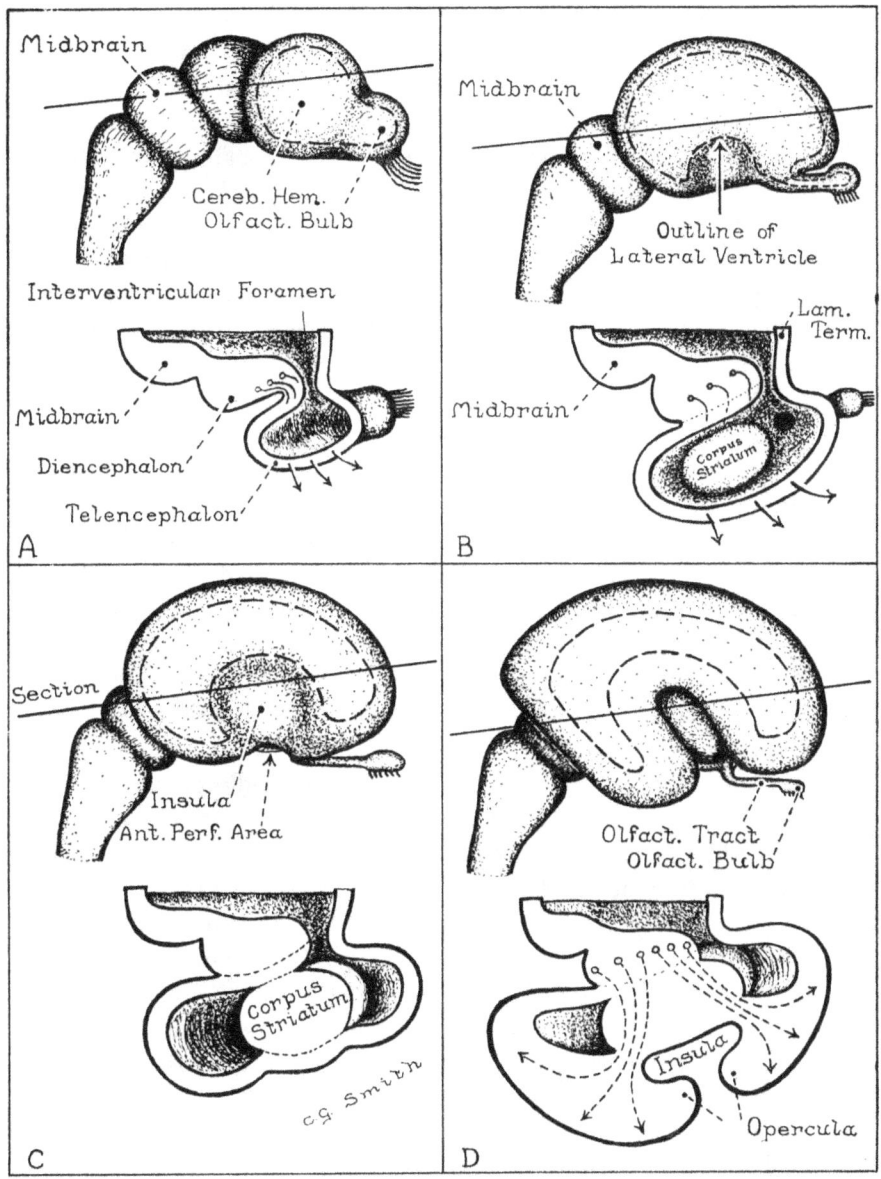

FIGURE 83. The development of the cerebral hemisphere. The lateral aspect and a horizontal section of each of four specimens illustrating stages in the development of the cerebral hemisphere.

with the enlarging hemisphere becomes the stalk of the olfactory bulb. This stalk is reduced, later on, to a bundle of nerve fibres and is called the olfactory tract. Thus development accounts for the attachment of the olfactory bulb to the cerebral hemisphere.

Let us now trace the growth of the attachment of the hemisphere to the diencephalon. Turn to figure 83A and note that all the fibres that connect the diencephalon and the cerebral hemisphere must pass through the caudal wall of the interventricular foramen. This connection obviously must grow bigger to accommodate the pathways that seek (a) to enter the cerebral hemisphere from the brain stem and (b) to descend from the cerebral hemisphere to the brain stem. Hence the "posterior" wall of the interventricular foramen grows progressively thicker and in so doing its caudal border shifts toward the midbrain along the lateral aspect of the diencephalon. The end stage is shown in figure 83B. In this way the cerebral hemisphere acquires an attachment to the whole lateral aspect of the diencephalon as well as to its rostral border. Note that the interventricular foramen is now at the anterior border of the much enlarged attachment of the cerebral hemisphere to the brain stem.

The next step in the development of the cerebral hemisphere is one in which the lateral ventricle changes from a more or less flattened, ovoid cavity to a C-shaped, tubular cavity. The formation of a C-shaped lateral ventricle is illustrated in diagrams 83 B, C, and D. In B the horizontal section shows a swelling in the middle portion of the floor of the lateral ventricle. This is produced by a mass of cell bodies and nerve fibres called the corpus striatum. In C the developing corpus striatum has become much larger and has spread across the floor to connect the inferior parts of the medial and lateral walls of the ventricle. In figure 83D the corpus striatum bulges upwards to occupy all of that part of the ventricle located lateral to the diencephalon. In so doing it converts the lateral ventricle into a curved tubular space extending rostrocaudally and outlining the rostral, the dorsal, and the caudal margins of the corpus striatum.

The corpus striatum, as it grows, not only encroaches on the lateral ventricle but also invades the floor of the interventricular foramen. In this way the ventral part of the communication between the lateral ventricle and the third ventricle is closed and, for a short time, the corpus striatum forms a part of the wall of the third ventricle. Later, the corpus striatum is excluded from the wall of the third ventricle by a layer of cells from the diencephalon that forms the preoptic region.

Let us turn now to a consideration of the outer aspect of the cerebral hemisphere. At the same time as the corpus striatum is developing, other nerve cells are accumulating on the external surface in a bark-like investment called the cerebral cortex. This covers all of the lateral surface of the hemisphere and all but a very small part of the medial and inferior surfaces (figure 84C). The cortex lateral to the corpus striatum behaves as if its growth were inhibited by this mass of grey matter. At any rate it fails to grow as rapidly as the surrounding cortex and as the hemisphere enlarges this area becomes the floor of an ever deepening depression on the lateral surface of the hemisphere (figures 83C and 83D). It has a more or less circular outline and is called the insula. The expanding marginal parts of the hemisphere progressively overlap the insular cortex behind, above, and in front. In the fully developed brain the overhanging parts meet along a straight line to form a fissure, the lateral fissure, open below and inclined obliquely upward and backward. This fissure may be likened to the palpebral fissure and the insula may be likened to the eyeball. Both the eyeball and the insula are covered by two lids. Those of the insula are called the opercula.

While the insula is developing on the lateral surface of the hemisphere the inferior and medial surfaces are also acquiring some of their features. Figure 84 shows the appearance of the inferior and medial surfaces about the middle of the fourth month of foetal life. Note that the olfactory bulb is attached by a slender stalk to the anterior border of a small, well defined area on the inferior surface of the cerebral hemisphere. This area has many small holes for blood-vessels and is called the anterior perforated area. It lies alongside the anterior half of the ventral (inferior) surface of the diencephalon and is bordered laterally by the cortex of the insula. It is a part of the inferior surface of the corpus striatum.

On the medial surface of the hemisphere two areas can be distinguished. These are (1) the septal area, significant because of its peculiar relationship to consciousness, and (2) the choroid membrane in which the choroid plexus of the lateral ventricle develops. The **septal area** is so called because part of it helps to form a median partition, the septum pellucidum, that develops later between the right and the left lateral ventricles. The characteristics that define the septal area are: (1) *its location*—it is a small part of the medial surface of the hemisphere that lies adjacent to, and is coextensive with, the lamina terminalis; (2) *its connections*—it receives the olfactory pathways that reach the medial aspect of the hemisphere; and (3) *its structure*—its cells are irregularly arranged to form poorly defined nuclei, not cortex.

The **choroid membrane** is a paper-thick portion of the medial wall of the lateral ventricle that remains thin because it is not invaded by

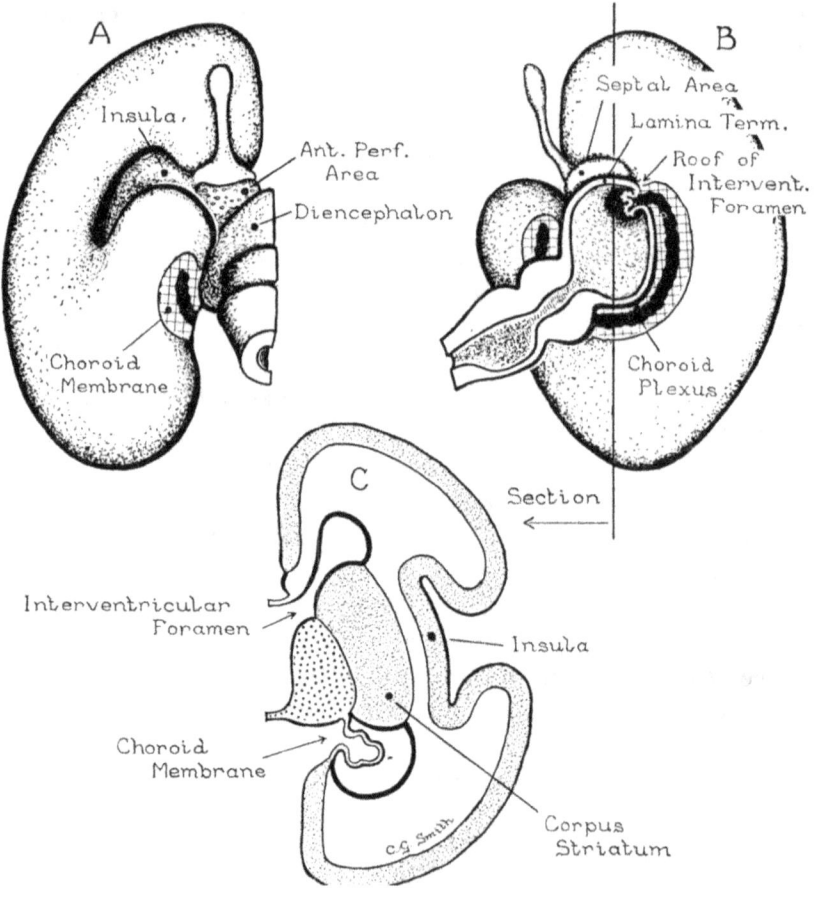

FIGURE 84. The right half of a foetal brain at 3½ months. A: Inferior aspect. B: Medial aspect. C: Horizontal section of the forebrain.

nerve cells (figure 84C). It is a ribbon-like band adjacent to and partially encircling the attachment of the hemisphere to the brain stem. It stretches from the interventricular foramen at its upper anterior end to a point behind the anterior perforated area. On the surface of this membrane and stretching its full length is a slender, cord-like mass of capillaries. These capillaries cause the membrane to bulge into the ventricle and thus is formed the choroid plexus of the lateral ventricle. The choroid plexus is prolonged onto the roof of the third ventricle at the interventricular foramen and is continuous with the plexus located there.

The differentiation of the septal region and the choroid membrane sets the stage for the development of the commissures that unite the cerebral hemispheres. To understand the development of the commissures it is important to appreciate that the first commissure to unite the hemispheres is a cellular one. The cells of this grey commissure are derived from the septal area. During the fourth month of foetal life cell masses from the right and the left septal areas invade that portion of the lamina terminalis that forms the anterior wall of the interventricular foramen and meet there at the midline. This cellular, thickened part of the lamina terminalis is visible from above as a transverse ridge located just in front of the membranous roof of the interventricular foramen (figure 85A). It provides a bridge of nervous tissue through which nerve fibres from one hemisphere can cross the midline to enter the other hemisphere. Since the first fibres to develop come from the portion of the hemisphere located lateral to the lamina terminalis they are added to the crest of this ridge and its height increases accordingly. Later, as fibres from successively more posterior parts of the hemisphere develop, they are added to the posterior border of the crest and the commissural band begins to grow back like a canopy for the diencephalon (figure 85B). This enlargement of the commissural band continues until it covers the upper surface of the diencephalon completely. The space between the diencephalon and the canopy is the transverse fissure of the cerebrum. It is of course open behind and closed in front and on each side. The last part of the cerebral hemisphere to mature is the frontal lobe, that is, the anterior portion. As it develops, its commissural fibres are added to the front of the commissure (figure 85C) and thus it acquires a triangular outline as seen in a midsagittal section.

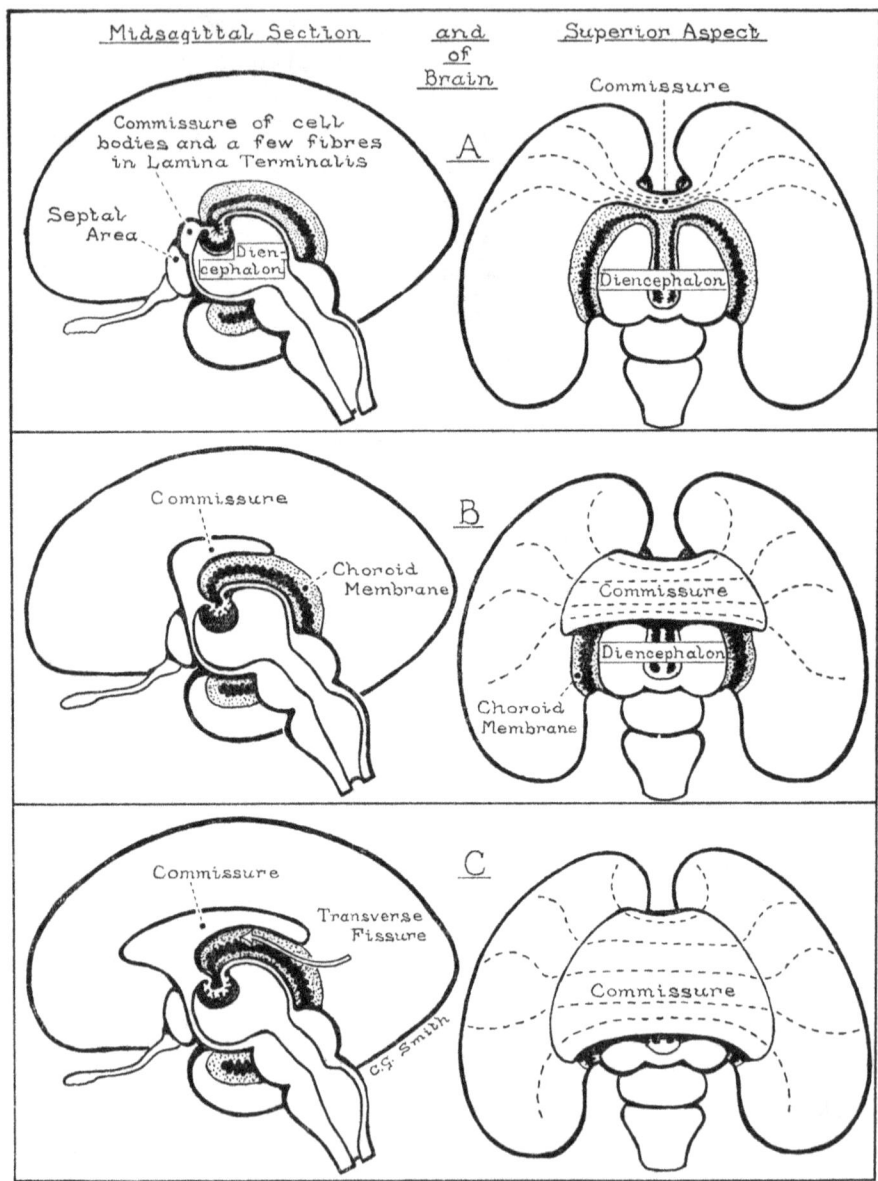

FIGURE 85. The first three stages in the development of the commissures that connect the cerebral hemispheres.

While the commissural band is growing to assume the size and shape pictured in figure 85C, the commissural fibres it contains are grouping themselves to form three distinct bundles as shown in figure 86. The largest of these is the corpus callosum; the other two are the hippocampal commissure and the anterior commissure.

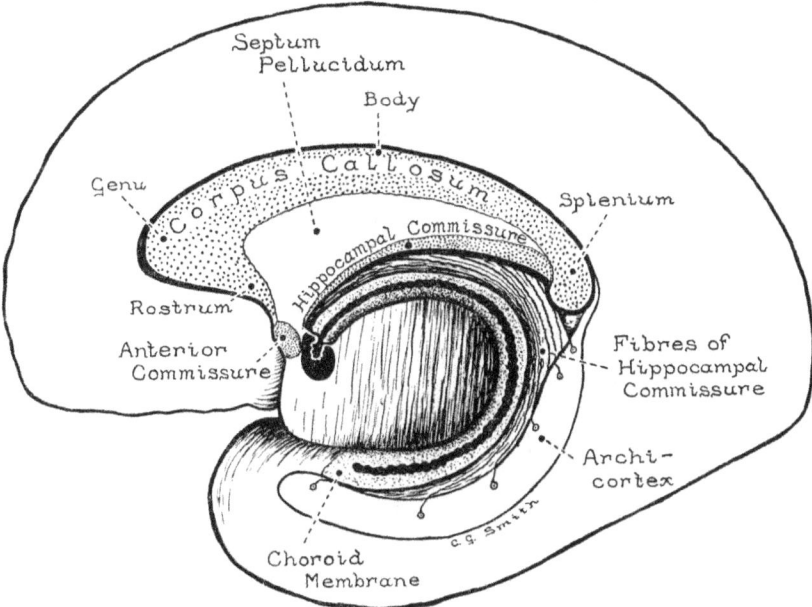

FIGURE 86. The medial aspect of the right cerebral hemisphere showing the fourth and final stage in the development of its commissures.

The **corpus callosum** is made up of fibres that connect the neocortex, that is, the newer cortex of the two hemispheres. These fibres form a thick, compact layer on the superior and on the anterior-inferior surface of the commissural band. The midsagittal section of the corpus callosum has four subdivisions. The bulbous posterior extremity is the *splenium* (bandage). The anterior extremity is the *genu* (knee). The intervening part is the *body*. The fourth part, the *rostrum* (beak) is the tapered portion that extends from the genu toward the lamina terminalis.

The **hippocampal commissure** connects the relatively very small areas of archicortex in the inferior parts of the right and left hemispheres. It forms therefore only a thin layer of fibres in the posterior-inferior border of the canopy-like mass connecting the two hemispheres.

The **anterior commissure** is a compact bundle, 2 or 3 mm. in diameter, located just in front of the interventricular foramen where the first fibres to unite the hemispheres crossed the midline. It contains some fibres that connect the right and left olfactory bulbs, but most of

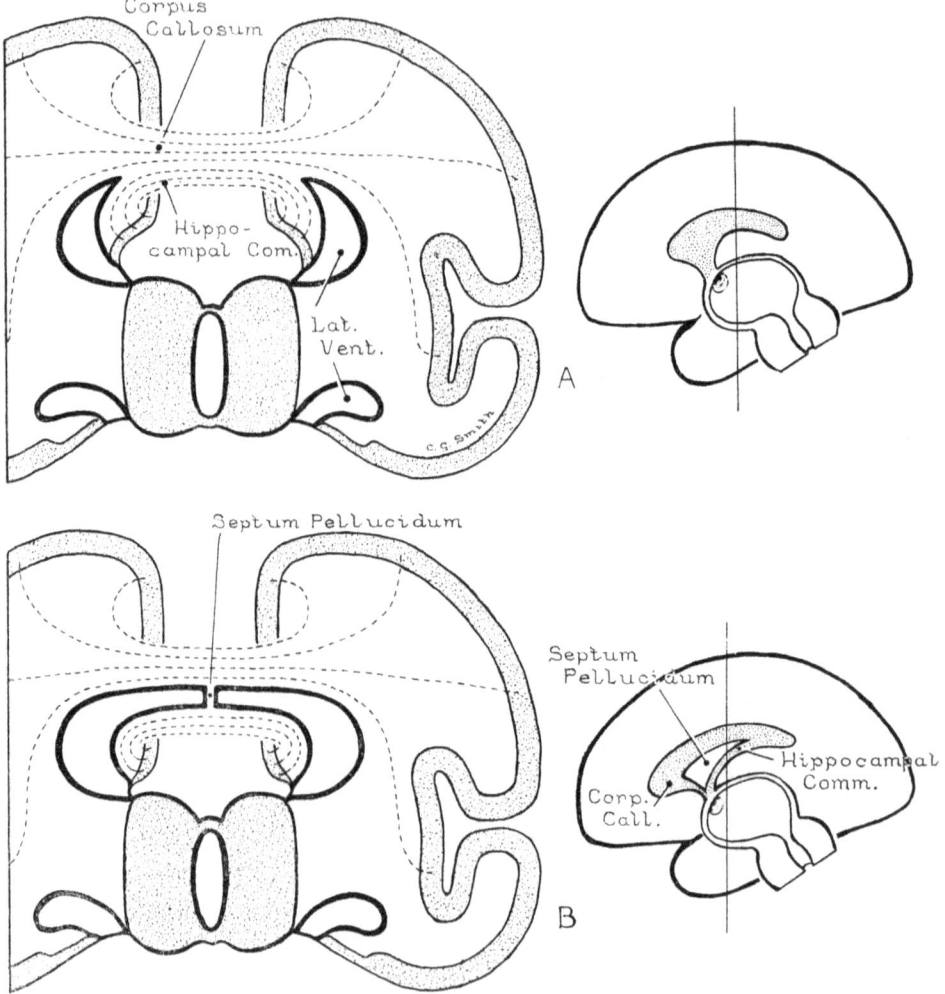

FIGURE 87. Frontal sections of the forebrain showing how the commissural fibre band is split by the right and left lateral ventricles to form the corpus callosum, the hippocampal commissure, and the septum pellucidum.

its fibres connect neocortical areas located in the inferior parts of the hemispheres (temporal lobes).

The crowding together of the commissural fibres at the surfaces of the canopy-like band uniting the hemispheres leaves the nerve cells it contains in a central location. These, the nerve cells of the original grey commissure uniting the right and left septal areas, are crowded together in turn by the expansion of the right and left ventricles. As illustrated by the drawings in figure 87, the right and left lateral ventricles enlarge and extend medially into the commissural band until they are separated only by a thin partition of nerve cells—the **septum pellucidum**. This partition is perforated in about 8 per cent of normal brains and in that case fluid can flow from one lateral ventricle to the other and be drained by either the right or the left interventricular foramen.

II. THE FORM OF THE CEREBRAL HEMISPHERE

The cerebral hemispheres plus the diencephalon to which they are attached fill a chamber that has a spherical, dome-like roof and a terraced floor (figure 88). The anterior part of the floor is the roof of the orbit. The posterior part of the floor, formed by the middle cranial fossa and the tentorium cerebelli, is at the level of the floor of the orbit. The two hemispheres occupy all the space available in this chamber. They project in front, above, and behind the diencephalon, extend medially to enclose it, and flatten out against each other at the midline. Thus each hemisphere acquires the shape of a quarter of a sphere with two notches—one for the diencephalon and one for the orbit. Its three surfaces face laterally, medially, and inferiorly respectively.

A. THE LATERAL SURFACE

This surface is part of a sphere and has a superior semicircular border that extends from the frontal pole—the extremity above the orbit—to the occipital pole, its posterior extremity (figure 88). The inferior border completes the semicircular outline but it is divided into a shorter anterior part above the roof of the orbit and a posterior part that is at the level of the floor of the orbit. The rounded, anterior end of the inferior portion of the hemisphere behind the orbit is the temporal pole.

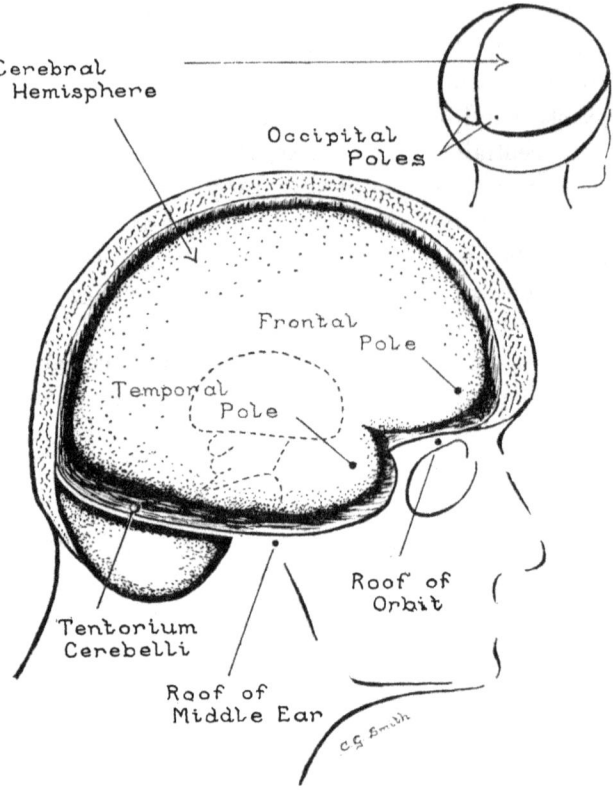

FIGURE 88. The right cerebral hemisphere exposed within the skull to show its lateral aspect and its chief relationships.

B. THE MEDIAL SURFACE

This surface is flat and its outline is that of a semicircle (figure 89). Just above the middle of its straight inferior border is the deep notch which is occupied by the diencephalon. The connection with diencephalon is in the floor of this notch, and in a sense this is a hilum of the hemisphere because through it pass all the pathways that connect the hemisphere and the brain stem. The upper wall of this notch—for the diencephalon—is separated from the superior surface of the diencephalon by the transverse fissure. Above the notch is the arc-like sagittal section of the commissural band that connects the right and left cerebral hemispheres.

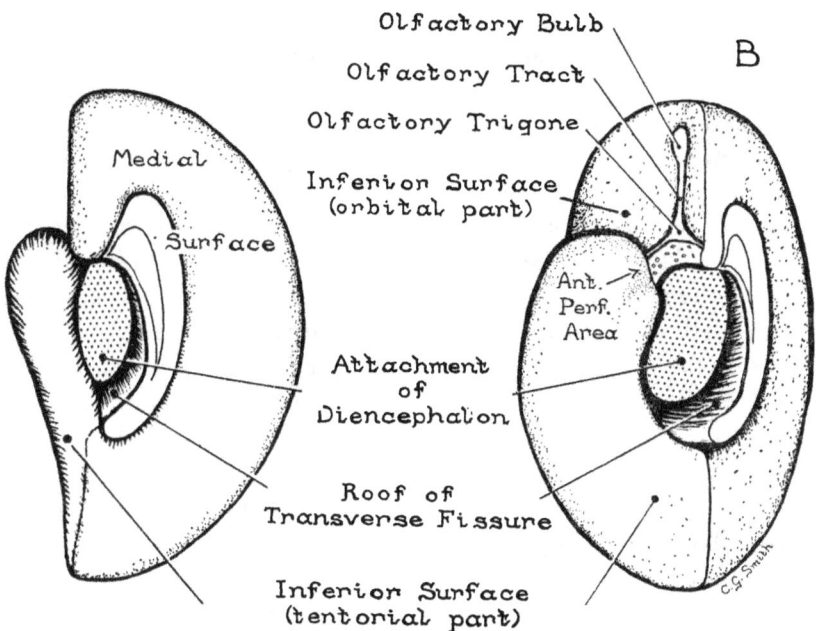

FIGURE 89. A: The medial aspect of the cerebral hemisphere. B: The inferior aspect of the cerebral hemisphere.

C. THE INFERIOR SURFACE

This surface has a lateral border that is a semicircle extending from the frontal pole to the occipital pole (figure 89). Its medial border completes the semicircular outline. At the middle of its straight medial border is the notch occupied by the diencephalon. The floor and walls of this notch were described with the medial surface. The anterior third of the inferior surface is in front of the notch for the diencephalon and rests on the orbit. The posterior two-thirds is on a lower level and is described as the tentorial surface although part of it rests on the floor of the middle cranial fossa. Neither the orbital nor the tentorial surface is horizontal. The orbital surface faces laterally as well as inferiorly and the tentorial surface faces medially as well as inferiorly. Hence the tentorial surface is usually described with the medial surface.

A feature of the inferior surface is the attachment of the olfactory bulb and the pathways leading from it to the cerebral cortex, corpus striatum, and septal region. The **olfactory bulb** is about the size of an orange seed and is connected to the hemisphere by a slender fibre bundle, the **olfactory tract**. This tract is attached to the apex of a small triangular area on the inferior surface of the hemisphere called the **olfactory trigone**. The latter is located just in front of the anterior perforated area, that is, about 1 cm. in front of the notch for the diencephalon, and about 1 cm. from the midline. The olfactory trigone serves as a distributing centre for the olfactory impulses. Pathways fan out into the anterior perforated substance and along its borders. The medial pathways course into the septal area, the lateral pathways course to the temporal pole. The fibres that follow the lateral border of the anterior perforated area are called the lateral olfactory stria. This stria is on the surface of a band of grey matter, the lateral olfactory gyrus, which is not demarcated from the cortex of the insula in man. Hence this part of the lateral olfactory gyrus is named as part of the insula, that is the **limen insulae** (edge or border of the insula).

III. THE CEREBRAL CORTEX

A. GENERAL CHARACTERISTICS

The cerebral cortex is a layer of grey matter 2 to 4 mm. thick that forms a bark-like covering for most of the hemisphere. In it occur the complex associations that are the function of the cerebral hemisphere. The only parts of the free surface that are not covered by cortex are the choroid membrane, which contains no nerve cells, the septal area which contains cells in irregularly arranged clusters, and the anterior perforated area which is a part of the corpus striatum.

Structurally, in addition to being a thin, superficial layer of grey matter, the cortex is peculiar in that its cells are arranged in three to six layers that are parallel to the surface. Functionally it may be considered to consist of unit masses that are microscopic, elongated blocks reaching from its outer to its inner surface. Since the cell content of each of the three to six layers in a unit can only vary within relatively narrow limits, it follows that the length of a unit and hence the thickness of the cortex is also relatively constant. This explains why the thickness of the cortex in man is only a little greater than in the rat.

Because the cortex is made up of cylindrical units set side by side to form a superficial layer of grey matter it follows that as more units are added more surface must be provided. To provide the very large surface that is required to accommodate the cerebral cortex of the human brain the surface of the hemisphere is wrinkled. Each furrow is called a **sulcus** and each ridge is called a **gyrus**.

B. THE CHARACTERISTICS OF ARCHICORTEX AND NEOCORTEX

Two very different kinds of cortex are found on the surface of the cerebral hemisphere; these are the archicortex, which has three layers, and the neocortex, which has six. The archicortex is present in reptiles and birds. In mammals the need for a more efficient association mechanism led to the development of the six-layered neocortex. This neocortex has proven to be so useful that it and not the archicortex has increased in amount with the need for a larger and more complex association mechanism. Hence in man the archicortex is only a small fraction of the total amount of cortex.

The histology of archi- and neocortex is illustrated schematically in figure 90. The simpler archicortex has a superficial, thin, almost acellular layer which receives the terminals of the fibres that enter the cortex. Deep to it are two layers of cells. These cells have fibres that convey impulses out of the cortex. Layer 1 of the archicortex is a layer of synapses. The afferent fibres of the cortex synapse here with the dendrites that extend into it from the cells in layers 2 and 3. In a histological preparation the sections of the dendrites and nerve fibres of layer 1 are visible as dust-like particles, hence layer 1 is called the molecular layer. Layer 2 is composed of pyramidal shaped cells. Each cell has a long apical strand of cytoplasm that ends, as already stated, in layer 1. Layer 3 is the polymorphic layer; its cells vary in size and shape and some have dendrites that end in layer 1. Both layers 2 and 3 contain some small cells whose axons do not leave the cortex. These provide extra indirect connections between the afferents of the cortex and the efferent cells of layers 2 and 3.

The neocortex is double the thickness of the archicortex. The increase in thickness is due to the addition of a layer of small pyramids sandwiched between two layers of small, round cells called granules. These three new layers are insinuated between the layer of large pyramids and the superficial molecular layer of archicortex. The afferent fibres come into the cortex as in the archicortex and push through to reach the surface. The dendrites of the large pyramidal cells also lengthen

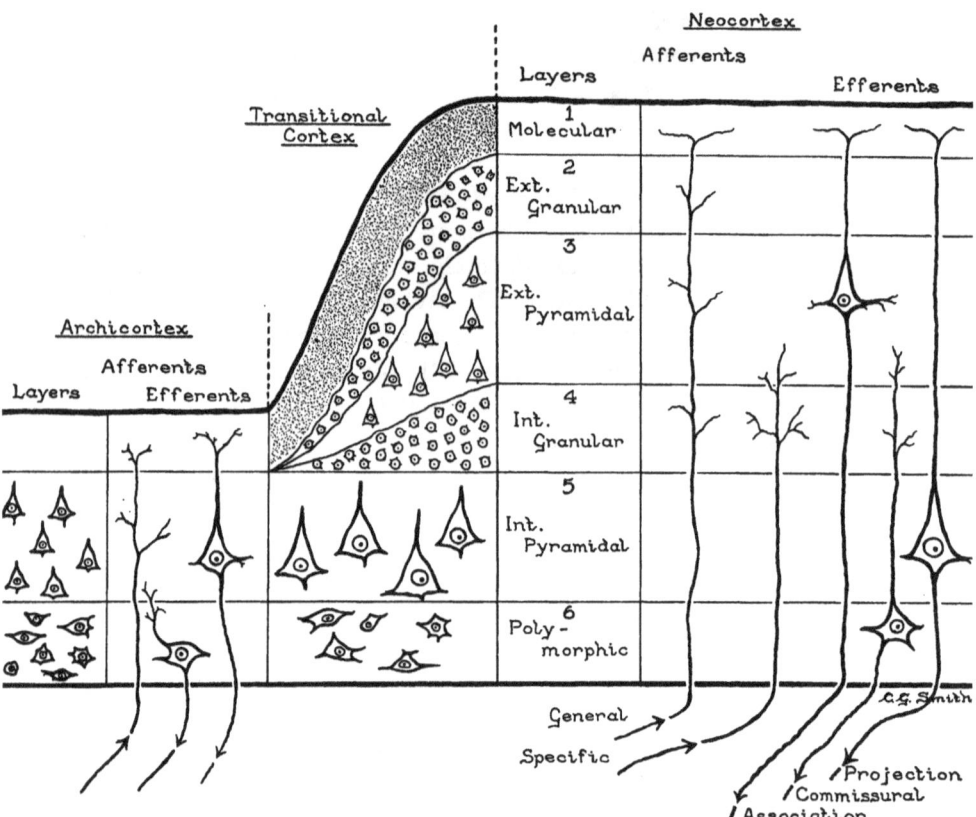

FIGURE 90. A diagram to show (a) the relationship of the layers of the neocortex to those of the archicortex and (b) the termination of the afferent fibres and the origin of the efferent fibres in each.

to reach the surface. In this way the molecular layer retains its superficial position. The afferent fibres of the cortex, however, have terminals ending not only in layer 1 but also in each of the new layers numbered 2, 3, and 4. In this way a large number of cells are interposed as indirect links between the afferent fibres and the efferent cells.

With the addition of more layers of cells in the superficial receiving portion of the cortical unit, the terminals of the sensory fibres (auditory, visual, touch, etc.) become more complicated and end in bush-like tufts. These endings distinguish them from other incoming fibres, e.g. commissural, and they are called specific afferents. The specific afferents end chiefly in layer 4 but also in layer 3. The larger branches of the bushy tuft of each fibre are myelinated and therefore in a histological preparation stained for myelin there is a thin layer of fine myelinated fibres in layer 4. In the visual area of the cortex, these myelinated branches are present in layer 4 in such numbers that they form a band that is visible in an unstained section as a white line. This is the **Line of Gennari.**

The addition of three layers of cells to archicortex to form the neocortex provides the cortex with additional cells for the reception and dispersal of impulses. The dispersal of impulses to other units of the same and opposite hemisphere is one of the functions of the cells of the external pyramidal layer. Cells with this function, however, have also been acquired by layers 5 and 6 of the neocortex.

In an attempt to summarize what has been said concerning the structure and function of the cortex we may state that in a general way impulses coming to the cortex are received in the superficial layers and impulses leaving the cortex leave from the deeper layers. The cells interposed between the terminals of the afferent fibres and the efferent cells are largely in layers 2, 3, and 4, but all layers have cells that serve this purpose. The processes of these cells do not leave the cortex. Their branching nerve fibres are so distributed that the cells of the cortex are linked to form self-exciting circuits within which impulses may go round again and again. Off-shoots of such a circuit may thus excite adjacent cells periodically. This is one explanation that has been offered for the rhythmic changes in surface potential of the brain at rest.

C. THE CHARACTERISTICS OF THE SENSORY, THE ASSOCIATION, AND THE MOTOR AREAS OF THE CORTEX

The cortex is a specialized association mechanism but some parts of this thin shell of grey matter have to serve as receiving stations for the incoming sensory pathways and one part has to serve as a dispatching centre, that is, as a site of departure for the long pathways that carry the messages to the muscles.

The receiving areas are the **sensory projection areas.** There are four of these, one for each of the known sensory pathways, namely, the optic, auditory, general sensory, and olfactory. These receiving stations are widely separated. Between them and also at a site in front of all of them we find the association areas proper. These **association areas** receive impulses from the sensory areas and do the essential work of processing the data from the environment. The willed response is initiated by impulses that spread from the association areas to the centrally located area of cortex called the **motor area**. This is the one that is connected with the motor nuclei of the brain and the cord. The descending pathways are the motor pathways. Of these the corticobulbar and the corticospinal, already studied in the brain stem, extend directly to the motor nuclei or at least to their immediate vicinity.

These cortical areas, of course, all look alike from the surface and cannot be identified by their appearance. They do, however, differ somewhat in histological structure. Thus, in sensory areas—the site of the termination of large numbers of afferent fibres—there is an increase in the number of granular cells in layers 2 and 4 and a corresponding decrease in the number of medium and large pyramids in layers 3 and 5. The replacement of large cells by small, granular cells leads to a marked reduction in the thickness of the cortex. The cortex of sensory areas is therefore thinner than that of other areas and because of the increased number of granular cells is called **granular cortex**.

The histological changes in the motor area are the counterpart of those in the sensory area. Here the need is for pyramidal cells with long nerve fibres to carry impulses to the brain stem. These cells are provided at the expense of the small, granular cells and the result is a much thicker cortex, about 4 mm. Because it has so few granular cells it is described as **agranular cortex**. All other areas are association areas and have the structure described above as typical of neocortex.

The small differences in histological structure outlined above are great enough to influence the location of the lines along which the cortex buckles as it spreads out to cover more surface. In some cases it buckles at the margin of an area (general sensory area), in other cases it buckles along a line across the middle of the area (visual area). Since the first parts of the cortex to be differentiated are the sensory and the motor areas, it follows that the first sulci to develop appear in their vicinity. These sulci then proceed to deepen as the developing association cortex crowds these areas. Hence, although obscured by secondary foldings of the cortex, they can in most cases be identified by their depth. This is possible in the dissecting room and, by reference to these deepest sulci, the cortical areas can be charted.

11. The Cerebral Hemisphere II: The Sulci, Gyri, and Cortical Areas

I. THE SULCI OF THE LATERAL SURFACE

A. SULCI USED IN DEMARCATING LOBES

THE LATERAL SULCUS

THIS IS the deepest sulcus of the lateral surface (figure 91). It has the insula for its floor and the opercula of the insula for its walls. The sulcus begins at the posterior border of the orbital surface where the frontal and the temporal opercula come together to cover the lateral part of the anterior perforated area. From here the sulcus extends backwards parallel to the inferior margin for about 2½ inches and then ascends a short distance to end near the centre of the lateral surface of the hemisphere.

THE CENTRAL SULCUS

The central sulcus is one of the most significant because it has the vitally important motor area in its anterior wall. It ascends almost

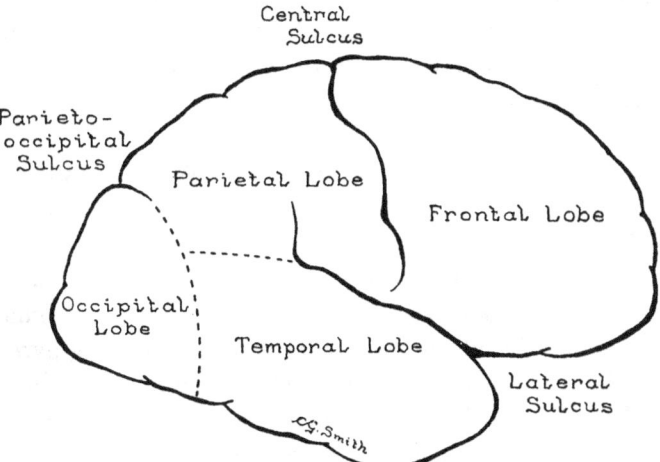

FIGURE 91. The lobes of the cerebral hemisphere.

vertically from a point immediately above the midpoint of the lateral sulcus to the midpoint of the superior border. It usually cuts across the superior border.

THE PARIETO-OCCIPITAL SULCUS

This is a deep sulcus of the medial surface that extends a short distance onto the lateral surface. It cuts the superior border of the hemisphere at a point in line with the lateral sulcus and extends 2 cm. toward it.

The cerebral hemisphere is divided into regions called **lobes** as shown in figure 91. The frontal lobe lies in front of the central sulcus. The occipital lies behind a line drawn parallel to the central sulcus, from the parieto-occipital sulcus to the inferior border of the hemisphere. The region between the frontal and the occipital lobes is divided into upper and lower portions, the parietal and the temporal lobes respectively. They are separated by the lateral sulcus and a line that extends this sulcus to the occipital lobe.

B. THE SULCI OF THE FRONTAL LOBE

THE PRECENTRAL SULCUS

This is parallel to the central sulcus and about a finger's breadth in front of it (figure 92). It is usually in two parts, a superior and an inferior, which overlap somewhat with the inferior part lying in front. The precentral sulcus may cut into the lateral sulcus but it rarely cuts the superior border.

THE SUPERIOR AND INFERIOR FRONTAL SULCI

The portion of the frontal lobe in front of the precentral sulcus is divided into upper, middle, and inferior portions of equal width by two sulci that extend forward from the upper and lower parts of the precentral sulcus. These are the superior and inferior frontal sulci. They are commonly interrupted by many small, bridging gyri and do not reach the anterior border of the hemisphere. An intermediate sulcus of variable depth and length is usually present between them.

Complicating the pattern of deep sulci of the frontal lobe but too variable to be useful as a guide to cortical areas are the sulci that cut into the frontal lobe from the lateral sulcus. These sulci are the result

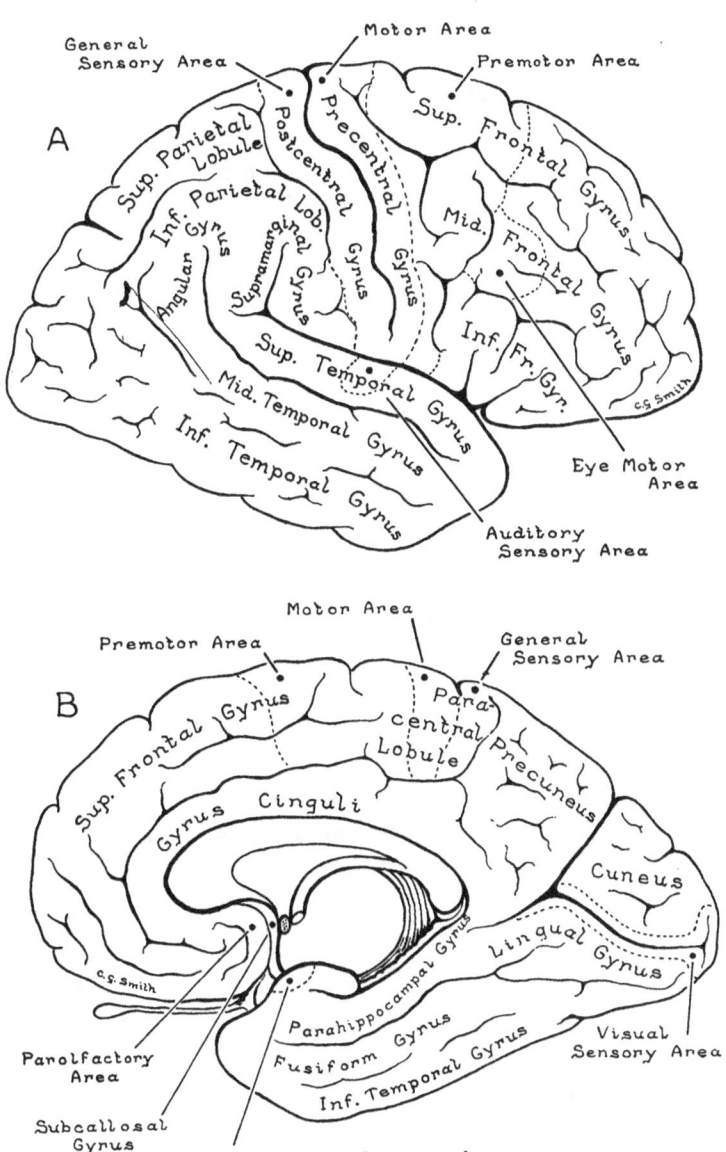

FIGURE 92. The gyri and the sensory and motor areas of the cerebral hemisphere. A: The lateral aspect. B: The medial aspect.

of an irregular buckling of the anterior operculum as it grows backwards over the insula.

C. THE SULCI OF THE PARIETAL LOBE

THE POSTCENTRAL SULCUS

The postcentral sulcus is parallel to the central sulcus and a finger's breadth from it (figure 92). It is usually divided into upper and lower parts and it stops short of both the superior border and the lateral sulcus.

THE INTRAPARIETAL SULCUS

This extends anteroposteriorly to divide the parietal lobe into a smaller (upper) and a larger (lower) portion. It usually is continuous at its anterior end with the inferior portion of the postcentral sulcus (phylogenetically the latter is a ventrally deflected part of the intraparietal sulcus). At its posterior end it enters the occipital lobe to end just behind the parieto-occipital sulcus.

D. THE SULCI OF THE TEMPORAL LOBE

THE SUPERIOR TEMPORAL SULCUS

This is parallel to the lateral sulcus and about a finger's breadth from it (figure 92). It begins near the temporal pole and ends by turning upwards into the parietal lobe as does the lateral sulcus.

THE MIDDLE TEMPORAL SULCUS

This is parallel to the superior temporal sulcus and a finger's breadth below it. It is not a continuous sulcus.

II. THE SULCI OF THE MEDIAL AND THE INFERIOR SURFACES

The evolution of the sulci of the medial surface and the tentorial portion of the inferior surface provides us with a key to their significance. As a result of the development of the large commissures connecting the hemispheres, the cortex of the smooth medial surface is crowded toward the outer border of the hemisphere. This causes it to buckle along a curved line extending anteroposteriorly. The sulcus so formed begins in front of the corpus callosum and arches back above it, and then downwards behind it, as in figure 93A. The fold of cortex

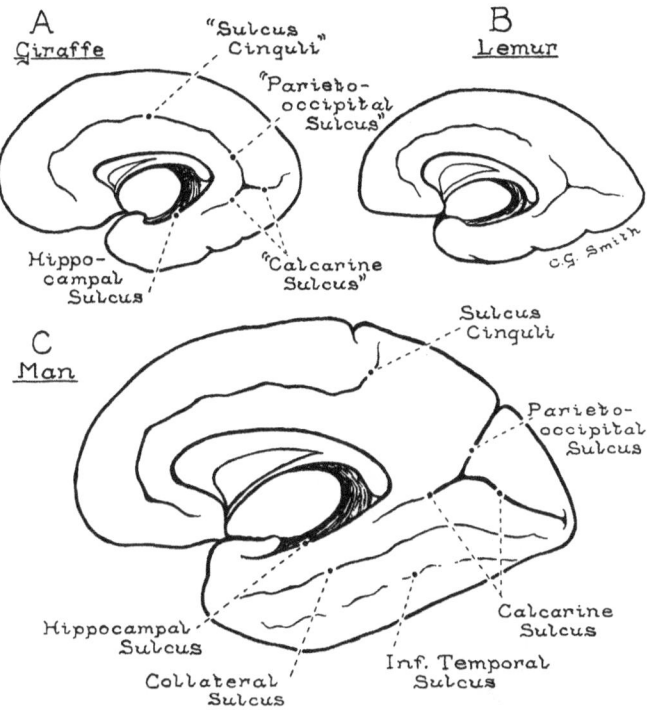

FIGURE 93. The evolution of the deep sulci of the medial and the tentorial surfaces of the cerebral hemisphere.

outside this sulcus is so sharply bent where it turns round the posterior border of the splenium that it in turn buckles to form a spur-like sulcus that extends back toward the occipital pole. Later, as the association cortex located above and a little in front of the splenium begins to expand, the floor of the ring-like sulcus at this site bulges to the surface again. Still later, as this region continues to expand, the contiguous ends of the now interrupted, ring-like sulcus are deflected toward the outer margin of the hemisphere. As they are separated more and more by the developing cortex between them, they also grow longer and deeper. The end result is shown in the drawing of the human brain (figure 93C).

At the same time as sulci are developing above the corpus callosum others are developing on the tentorial surface. The first one to appear is adjacent and parallel to the attachment of the hemisphere to the diencephalon. It extends from the posterior border of the corpus callosum to a point close to the temporal pole. The cortex in the floor of this sulcus is archicortex. This is only half as thick as the neocortex accumulating along its outer border and as it is crowded medially it buckles and develops a furrow along its long axis. It is called the hippocampal sulcus (figures 93C and 94).

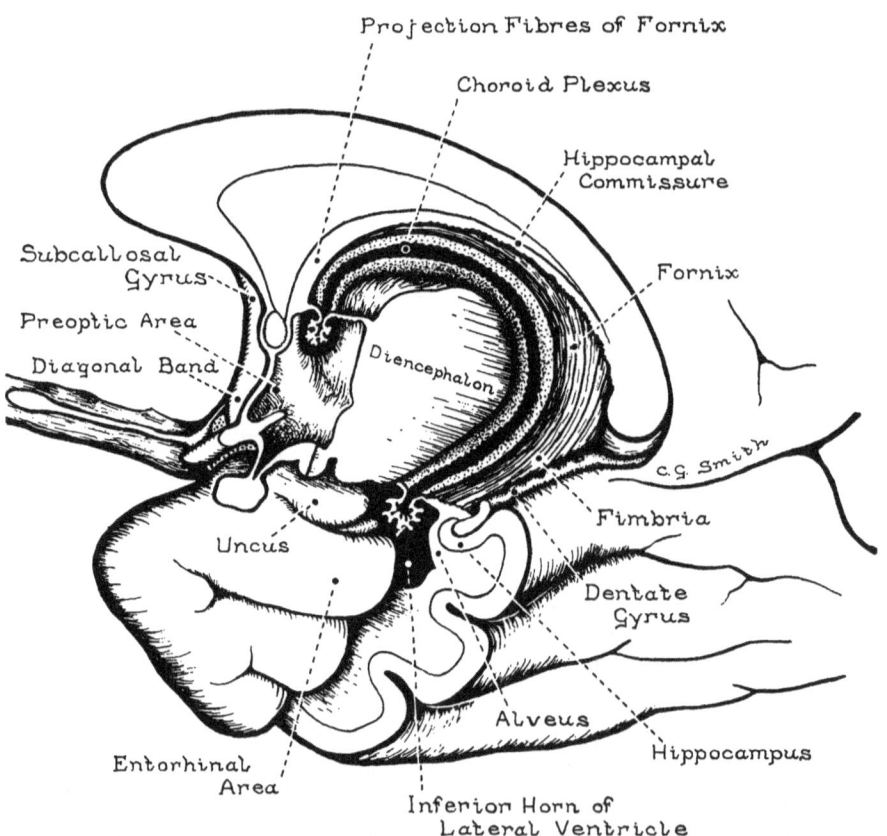

FIGURE 94. The medial aspect of the temporal lobe showing the parts of the hippocampal formation and its efferent pathway.

With the further growth of the neocortex on the tentorial surface, two more longitudinal sulci appear one after the other in progressively more lateral locations. The fully developed pattern of sulci is illustrated in figure 92B. Let us name and briefly describe them.

THE SULCUS CINGULI

This begins beneath the rostrum of the corpus callosum and arches upwards and backwards a finger's breadth from the margin of the cerebral hemisphere to reach a point just above the splenium. Here it turns toward the superior border of the hemisphere to end without reaching it, just behind the central sulcus.

THE CALCARINE SULCUS

This begins below the splenium and extends to the occipital pole. It has anterior and posterior parts of equal length that meet at a right angle at a high point of 1 cm. above the inferior border of the hemisphere. The anterior part of the calcarine sulcus is a detached portion of the cingulate sulcus. The posterior part is an offshoot of the sulcus that crosses the middle of the visual sensory area, to be described later.

THE PARIETO-OCCIPITAL SULCUS

This extends from the junction of the anterior and posterior parts of the calcarine sulcus to the superior border. It reaches the superior border about 5 cm. from the occipital pole. The anterior part of the calcarine sulcus and the parieto-occipital are in line and together are the detached portion of the sulcus cinguli.

THE HIPPOCAMPAL SULCUS

This sulcus is at the medial border of the tentorial surface and is located in the cleft between the cerebral hemisphere and the diencephalon. It lies alongside the membrane containing the choroid plexus of the lateral ventricle and can be identified by the nodular ridge of grey matter that lies in its floor (see figure 94). At its posterior end the hippocampal sulcus reaches the back of the splenium where it is continuous with the **callosal sulcus**, which outlines the corpus callosum. Anteriorly, beyond the end of the choroid membrane, the hippocampal sulcus deviates into the gyrus that lies along its lateral border and continues on for a short distance toward the temporal pole. The portion of the gyrus that lies medial to its anterior end is called the *uncus* (hook). This, *in situ*, is located along the medial border of the tentorium cerebelli and is in a position where it may be forced down between the tentorium and the midbrain when the upper portion of the brain swells or is pressed down by a tumor.

THE COLLATERAL SULCUS

This begins close to the occipital pole and extends almost to the temporal pole along a line that is a finger's breadth from the medial border of the tentorial surface. A shallow groove extends anteriorly and medially from a point near its anterior end to complete the outline of the uncus. This shallow groove is the equivalent of the rhinal sulcus in brains of lower animals. It marks the lateral border of the archicortex (figure 93C).

THE INFERIOR TEMPORAL SULCUS
(Occipito-temporal Sulcus)

This is a very irregular sulcus with many breaks. It is located lateral to the collateral sulcus and parallel with it.

III. THE GYRI OF THE CEREBRAL HEMISPHERE

The named gyri of the hemisphere are the broad ridges of cortex demarcated by the deep sulci (figure 92). In the frontal lobe there is a precentral gyrus and a superior, a middle, and an inferior frontal gyrus. In the parietal lobe there is a postcentral gyrus and behind this an upper and a lower gyrus separated by the intraparietal sulcus. These broad gyri are called the superior and inferior parietal lobules. The inferior parietal lobule has an anterior portion around the end of the lateral sulcus. This is the supramarginal gyrus. Behind it is the angular gyrus, which caps the end of the superior temporal sulcus. On the lateral surface of the temporal lobe there is a superior, a middle, and an inferior temporal gyrus.

On the medial surface of the hemisphere the gyrus cinguli lies between the sulcus cinguli and the corpus callosum. It is separated from the corpus callosum by the callosal sulcus except in front of the rostrum where there is an intervening smooth, narrow ridge called the subcallosal gyrus. This is not a part of the cortex; it is a part of the septal area of the hemisphere. It tapers to a point toward the genu but it is prolonged inferiorly along the anterior border of the lamina terminalis to become continuous with the anterior perforated area. The subcallosal gyrus, as already pointed out, is only a part of the septal area. The rest is in the septum pellucidum. This is usually connected with the subcallosal gyrus between the tip of the rostrum and the anterior commissure. It will be helpful to point out that the term septal region is used loosely for a more extensive region in front of the lamina terminalis that takes in a part of the cingulate gyrus and the superior frontal gyrus. The latter additional portion is usually demarcated by a shallow vertical sulcus and is called the parolfactory area.

Between the anterior part of the gyrus cinguli and the superior margin of the hemisphere we find the superior frontal gyrus. This is continuous posteriorly with the cortex around the end of the central sulcus. The latter cortex, called the paracentral lobule, is an extension of the cortex of the precentral and the postcentral gyri. The wedge-shaped part of the occipital lobe enclosed by the parieto-occipital and the calcarine sulci is the cuneus and the portion of the parietal lobe between the cuneus and the paracentral lobule is the precuneus.

On the tentorial portion of the inferior surface of the hemisphere there is a medial (parahippocampal), a lateral (inferior temporal) and an intermediate (fusiform) gyrus. The anterior end of the parahippocampal gyrus is bent back around the end of the hippocampal sulcus to form the uncus. The grey matter of the surface of the uncus appears to be an extension of the cortex of the parahippocampal gyrus but this is misleading because it is formed in part by the grey matter of the amygdaloid nucleus, a portion of the corpus striatum. The posterior end of the parahippocampal gyrus forks. One prong is continuous with the gyrus cinguli, the other is the lingual gyrus which extends back to the occipital pole between the calcarine sulcus above and the collateral sulcus below.

IV. THE SENSORY AREAS OF CORTEX

A. THE SENSORY AREAS OF NEOCORTEX

THE GENERAL SENSORY AREA

This is the area that receives the pathways from sense organs for touch, pain, temperature, and position. It is coextensive with the postcentral gyrus and extends into the paracentral lobule (figure 92) on the medial surface. It also extends onto the upper lip of the lateral sulcus. One might expect to find subdivisions of this area for touch, pain, etc., but this is not the case. It would appear that the pathways from all the different kinds of sense organs, that are posted at a given point in the body, end in the same area of cortex. These areas—each representing a part of the body—are arranged in an orderly fashion from the toes to the region of the mouth along the length of the postcentral gyrus. Areas for the toes and feet are in the paracentral lobule and those for the mouth region on the deep surface of the lip of the lateral sulcus. The alimentary canal has a representation in an extension of this general sensory area onto the insula.

THE AUDITORY SENSORY AREA

This area is coextensive with the anterior transverse temporal gyrus which is located on the lower wall of the lateral sulcus. This gyrus extends from the posterior-superior angle of the insula to the middle of the superior temporal gyrus. Only the end of the strip of auditory cortex reaches the lateral surface (figures 92 and 99).

THE VISUAL SENSORY AREA

This is an area which can be charted by a dissector with a hand-lens because on section the cortex is divided into superficial and deep layers by a fine, white line called the **Line of Gennari**. This is easily seen in unstained slices of the cortex. The cortex of the visual area is in both walls of the posterior part of the calcarine sulcus. Hence, only the borders of this area are visible on the medial surface. The upper half of the retina projects to the upper wall, the lower half to the lower wall and the macula is at the occipital pole.

B. THE SENSORY AREA OF ARCHICORTEX
THE OLFACTORY SENSORY AREA

This is the only sensory area of archicortex. It is located on the anteromedial part of the uncus (figure 92B). It differs from the neocortical sensory areas in having its afferent fibres enter its outer surface. This is because the olfactory tract is attached to the outer aspect of the hemisphere, whereas the other sensory pathways enter the hemisphere through the internal capsule.

V. THE MOTOR AREAS OF THE NEOCORTEX

A. THE CHARACTERISTICS OF PYRAMIDAL AND EXTRAPYRAMIDAL MOTOR AREAS

The **pyramidal motor areas** are the motor areas proper. They give rise to the descending fibres of the corticobulbar and corticospinal tracts. These tracts extend all the way to the immediate vicinity of the motor nuclei where their fibres either synapse with the motor cells or with internuncial cells connected with them by short fibres. These are the direct motor pathways of the cortex and they are designated pyramidal because most of them course through the pyramid on the surface of the medulla.

The cerebral cortex also has other areas called **extrapyramidal motor areas**. These give rise to fibres that are the beginning of multineuronal pathways to the muscles. The relay stations of these pathways, such as the subthalamic nucleus, the red nucleus, the substantia nigra, and the reticular nuclei, are in the corpus striatum and in the brain stem. No portion of these pathways is found in the pyramid, hence they are classified as extrapyramidal.

The motor areas proper have the following characteristics: (1) motor responses are elicited with weak stimuli, (2) each part of the body, for example a finger, can be activated independently, (3) removal of the whole motor area bilaterally is followed by paralysis of all voluntary movement.

The extrapyramidal motor areas differ from motor areas in that (1) responses are obtained only with strong stimuli, (2) the responses are a movement of a whole limb or of the head and eyes, (3) removal of the area is followed by awkwardness not paralysis.

B. THE LOCATION OF THE PYRAMIDAL AND THE EXTRAPYRAMIDAL MOTOR AREAS

The motor area that activates all parts of the body except the eyes is a strip of cortex that extends the length of the precentral gyrus. One border of this band of cortex is at the floor of the central sulcus, the other is near the crest of the precentral gyrus. Its medial end is in the paracentral lobule. Here stimulation excites movements of the foot and from this site to the region of the lateral sulcus, where the muscles of the head can be activated, the parts of the body are represented in order.

The motor area for eye movements is detached from the principal motor area. It is in the posterior end of the middle frontal gyrus.

Two areas of excitable cortex that qualify as extrapyramidal motor areas will be described here. They are the **premotor area** and the cortical area in the occipital lobe which controls the movements of the **eyes for near and far vision and in retaining a constant visual field.**

The premotor area is a band of cortex parallel to, and in front of, the motor area. It is about 5 cm. wide at the superior border of the hemisphere and about 1 cm. wide at the lateral sulcus. The extrapyramidal area of the occipital lobe controlling eye movements surrounds the visual sensory area. Its efferent fibres descend to the tectum of the midbrain.

C. FUNCTIONAL DEFICITS DUE TO LESIONS OF THE PYRAMIDAL AND THE EXTRAPYRAMIDAL PATHWAYS

Interruption of the motor pathways in the internal capsule is followed by paralysis of some muscles on the opposite side of the body. In the head, only the muscles of the lips and the genioglossus of the opposite side are paralyzed. In the limbs of the opposite side the muscles of the distal parts are paralyzed but those of the proximal part only partially and those of the trunk very slightly, if at all. The explanation for the retention of voluntary control of some muscles of the affected side when the pathway from one hemisphere is cut is the existence of an accessory pathway to these muscles from the homolateral hemisphere.

Interruption of the motor pathways in the internal capsule cuts not only the pyramidal fibres, such as the corticospinal, it cuts some significant extrapyramidal fibres as well. These accompany the corticospinal tract as far as the midbrain and then enter the tegmentum to descend to the reticular formation of the medulla. Here in the medulla is the origin of the reticulospinal tract that descends to the motor nuclei to modulate, that is, partially suppress, reflex responses. This suppression of reflex excitation is an important part of the mechanism of voluntary control of muscles. When it is lost the motor nuclei of the paralyzed muscles are exposed to the uncontrolled and more or less continuous excitation from such sense organs as the internal ear and the stretch receptors. Hence the paralyzed muscles are partially contracted at all times. Some muscles are more contracted than others and as a result the paralyzed half of the body assumes a characteristic posture. The paralysis is described as **spastic paralysis**.

Reflex responses in spastic paralysis are altered in a characteristic way. Stretch reflexes, such as the knee-jerk, are hyperactive but some, called superficial reflexes, are either lost or completely changed. The abdominal reflex is an example of a response that is lost. Stroking the skin of the abdomen on the paralyzed side fails to excite a local contraction of the underlying muscle as it does normally. A reflex that is completely changed is the one obtained by stroking the sole of the foot. Normally the toes curl as if to grasp the offending stimulus but after voluntary control of the foot is lost the response obtained is extension and fanning of the toes. The explanation for the decrease in superficial reflexes and the reversal of the response from the sole of the foot is not available. A recent discovery by Brodal of heretofore unsuspected fibres in the pyramid carrying impulses from sense organs directly to the motor cortex suggests that superficial reflexes are mediated by a long reflex arc that passes through the cerebral cortex.

VI. THE ASSOCIATION AREAS OF NEOCORTEX

It is convenient to classify association areas as primary, secondary, and tertiary.

Primary association areas are adjacent to the sensory areas. In these areas the data obtained by one of the senses are associated. For example, in the area adjacent to the visual area, the charcteristics of an object in the visual field such as colour, shape, surface features, will permit the recognition of an object like an orange.

Secondary association areas separate the primary association areas of cortex and receive their data from them. Hence in these areas, the data from several senses are associated to identify an object in the environment. It will be evident that the middle part of the cortex of the parietal lobe will be equidistant from the general sensory area, the visual area, and the auditory area and will therefore be in a position to be used in the most complicated association processes.

Tertiary association areas are located in the frontal lobe in front of the extrapyramidal area and the motor area for the eye. The development of this cortex is one of the characteristic features of the human brain. In lower animals such as the cat only a small amount of this cortex is present and consequently the motor area is close to the frontal pole. Tertiary association cortex is not required in the association process that leads to recognition of portions of the environment. It is not even required for the complex mental activity involved in speaking, reading, and writing. It would appear to be the site of higher forms of mental activity involved in solving problems. Removal of this cortex, which is usually called the prefrontal cortex, results chiefly in a loss of initiative.

Some evidence for the functional localization in the association cortex, as outlined above, is obtained in studies of the cortex in cases of aphasia. Aphasia is a diseased state in which speech is impaired. In reporting the following findings it is necessary to explain that the process of association by its very nature is diffuse and that a lesion in any part of the cortex will impair the whole mechanism of association to some extent. Hence the deficit reported for a localized lesion is only the most obvious one. The reported findings are as follows: (1) If a lesion is adjacent to the visual area, written words can be seen but not understood. (2) If a lesion is adjacent to the auditory area, spoken words can be heard but not understood. (3) If a lesion is in the inferior parietal lobule in the region of the angular gyrus there is a loss of ability to understand both written and spoken words. These lesions in, or at the border of, the parietal lobe impair the understanding of speech, and only secondarily disturb the ability to speak. The ability to speak would appear to depend upon an area in the frontal lobe in which the complicated motor activity involved in speaking is organized. This area is in the posterior part of the inferior frontal gyrus. A small lesion here may impair the ability to speak although no muscles are paralyzed and although understanding of speech is retained.

The study of aphasia as described above has revealed also that association areas of one hemisphere are dominant over those of the other hemisphere. For example, in right-handed individuals the control of speech is in the left hemisphere. Hence if the destruction of the association areas described above is to result in aphasia the lesions must be in the dominant hemisphere. Destruction of association areas in the non-dominant hemisphere leaves speech unimpaired.

VII. THE NEOCORTICAL AREAS IN TERMS OF THALAMIC CONNECTIONS

It is remarkable that each nucleus of the thalamus has an area of neocortex with which it is connected. This is a two-way connection so that impulses can be conveyed from thalamus to cortex and also from the cortex to thalamus. The two-way connection forms a self-exciting circuit. Figure 95 is a diagram showing the cortical areas connected

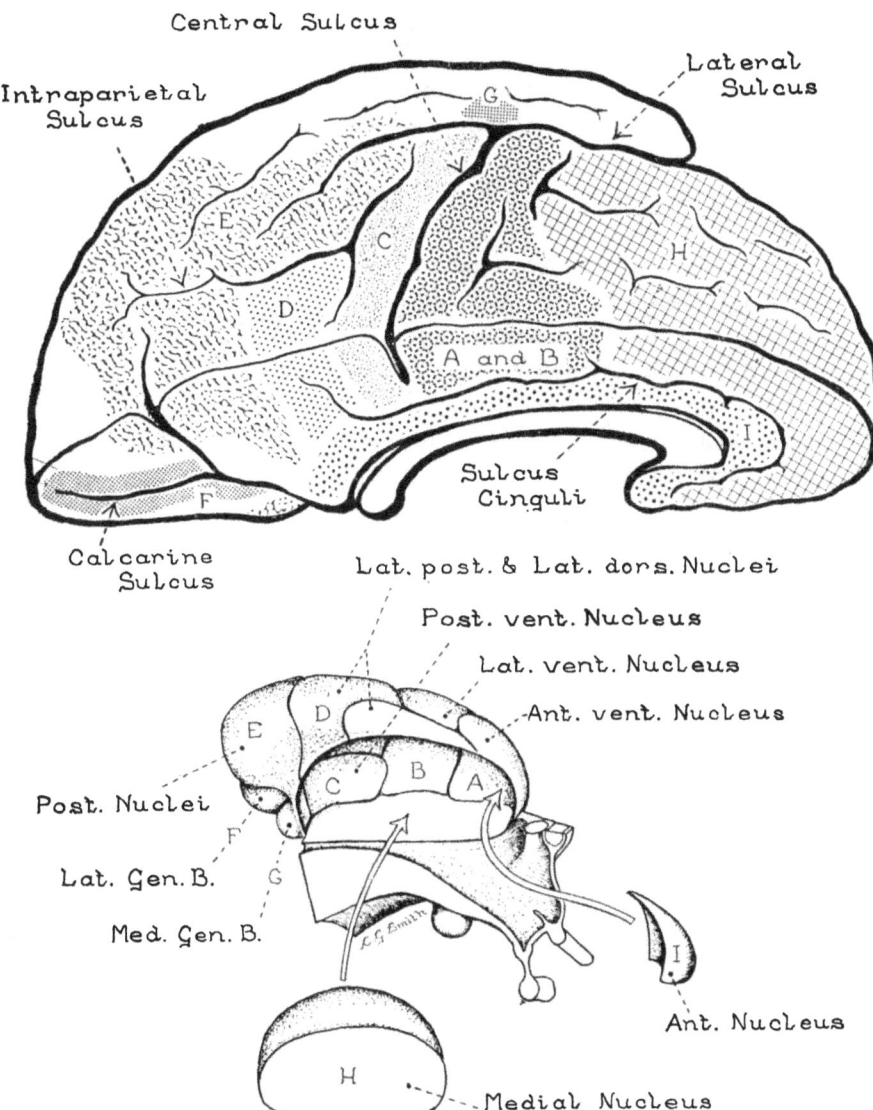

FIGURE 95. A diagram showing the cortical area connected with each of the thalamic nuclei. The corresponding parts of the thalamus and cortex are labelled with the same capital letter.

with each of the thalamic nuclei. Note that the only areas of cortex not connected with a nucleus of the thalamus are in the occipital lobe and in the temporal lobe. The area in the occipital lobe and a part of the area of the temporal lobe have connections with the tectum of the midbrain.

The discovery of the two-way cortical connections of each of the thalamic nuclei suggests that they may play an important part in the process of cortical association. For example, the two-way connection may serve as a circuit in which impulses can circulate for an appreciable time and prolong the excitation of a given area of cortex. Also it is possible to have widely separated cortical areas linked together through the thalamus. Such a linkage of the parietal cortex and the prefrontal cortex is illustrated in figure 96. Note that the existence of a two-way connection between the medial nucleus of the thalamus and the hypothalamus opens up the possibility that the cingulate gyrus, which receives a pathway from the hypothalamus, may be activated indirectly by cortical activity in the prefrontal cortex or in the parietal cortex. The pathway from the hypothalamus to the cingulate gyrus is by way of the mammillothalamic tract and the anterior nucleus of the thalamus.

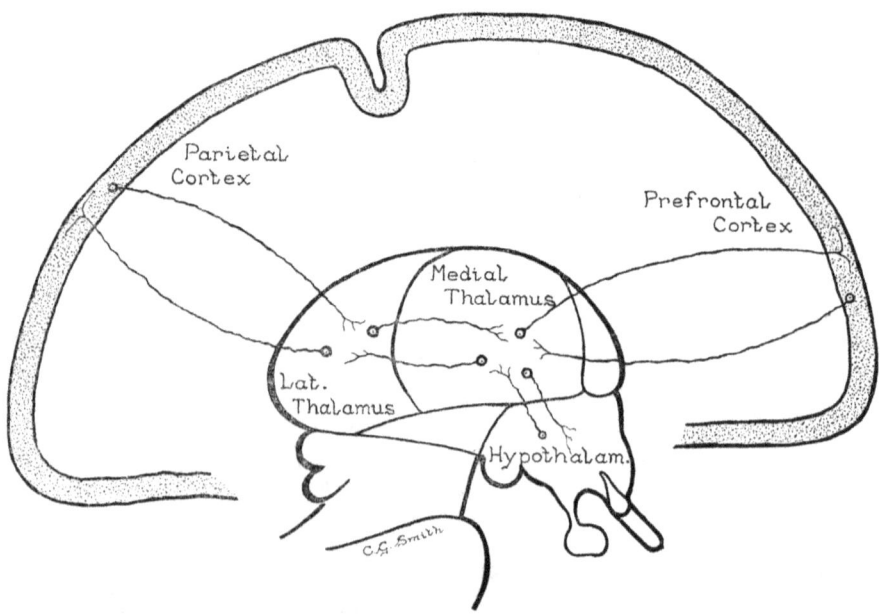

FIGURE 96. A diagram showing how the parietal cortex, the prefrontal cortex and the hypothalamus are connected by way of the thalamus.

VIII. THE ASSOCIATION AREAS OF THE ARCHICORTEX AND THE AREA OF ARCHICORTEX THAT GIVES RISE TO PATHWAYS DESCENDING INTO THE BRAIN STEM

The chief association area of archicortex is in the anterior half of the parahippocampal gyrus. This large area of cortex, called the entorhinal area, extends forward in the parahippocampal gyrus as far as the olfactory sensory area and it of course receives fibres from it. The fibres from the olfactory area are outnumbered, however, by fibres that enter this association cortex from the neocortex. The latter come from the adjacent neocortex of the tentorial surface and the more remote cortex of the medial aspect of the hemisphere. The influx of these association fibres from the neocortex has caused this area of archicortex to increase in size and to acquire additional cell layers. Hence its structure is transitional between archicortex and neocortex. This association cortex is connected by nerve fibres with two areas of typical three-layered cortex located in the wall of the hippocampal sulcus. These two areas are the hippocampus and the dentate gyrus and together they make up the hippocampal formation.

The cortex of the hippocampus is directly continuous with the entorhinal cortex and forms the wall of the hippocampal sulcus. This is the cortex that gives origin to the fibres of the fornix, the bundle that descends into the hypothalamus. The fornix is in certain respects comparable to the extrapyramidal pathways of the neocortex. It fails to qualify, however, as a motor pathway because stimulating it does not excite movements. The other part of the hippocampal formation is the dentate gyrus. This is a narrow band of association archicortex that differs from the hippocampal cortex in having smaller cells in its middle layer. It lies along the floor of the hippocampal sulcus adhering to the surface of the hippocampal cortex that forms its walls. It has a free surface, visible on separating the walls of the sulcus, with tooth-like elevations that account for its name. The posterior end of the dentate gyrus tapers and flattens out on the back of the splenium where it is continuous with a vestigial thin layer of cells on the surface of the corpus callosum called the *indusium griseum* (grey tunic). The dentate gyrus, as already stated, is an association area interposed between the entorhinal cortex and the hippocampus. Some of the fibres from the entorhinal area end in it, others bypass it and go directly to the hippocampus.

IX. THE OLFACTORY PATHWAYS THAT BYPASS THE CORTEX OF THE HEMISPHERE TO REACH THE BRAIN STEM

Since the cerebral hemisphere develops in the evaginated olfactory portion of the forebrain at a site between the attachment of the olfactory nerve and the brain stem, it follows that all olfactory pathways *must* enter the hemisphere. Some of them end in the cortex of the hemisphere (olfactory sensory area) but others pass through the hemisphere to reach the brain stem. These latter pathways bypass the cortex. Some of them have relay stations in the septal area, others in the anterior perforated substance, and still others in the amygdaloid nucleus. The pathways to the septal area are in the medial olfactory stria, those to the anterior perforated substance enter it from the olfactory trigone, and those to the amygdaloid nucleus are included with the cortical fibres in the lateral olfactory stria. From each of these relay stations fibres descend to the hypothalamus which is continuous with both the septal area and with the anterior perforated substance. The fibres from the amygdaloid nucleus may reach the hypothalamus through the anterior perforated substance or via the stria terminalis (to be described with the connections of the corpus striatum).

The septal area also gives rise to a pathway that serves as a back door to the archicortex. The fibres of cells in the subcallosal gyrus and the septum pellucidum penetrate the thickness of the corpus callosum and reach its superior surface. There they come together to form delicate, thread-like bundles, the longitudinal striae, that course back close to the midline and turn round the posterior end of the splenium to enter the dentate gyrus. This pathway conveys impulses, that have reached the septal area from the hypothalamus and the olfactory sense organ, to the association cortex of the hippocampal formation. It may be compared to the non-specific pathways that leave certain of the thalamic nuclei to reach neocortical association cortex.

X. THE LIMBIC LOBE

Until recent times it was believed that the hippocampal formation was activated chiefly by impulses from the sense organ of smell. This belief was based on observations in lower animals where the sense of smell and the size of the hippocampal formation appear to be correlated. Because man has a small olfactory sense organ and a large hippocampal formation, and because of evidence derived from clinical and pathological studies, Papez[1] was led to suggest that the hippocampal formation served as the efferent portion of a part of the hemisphere which functioned as the neural mechanism of emotion. The portion of the hemisphere included in this suggested mechanism of emotion is a C-shaped region on the medial surface of the hemisphere adjacent to and partially surrounding the attachment to the diencephalon. This is the limbic portion (*limbus*, edge, border) of the hemisphere and is known as the limbic lobe.

The limbic lobe includes the subcallosal gyrus, the parolfactory area, gyrus cinguli, parahippocampal gyrus, and the hippocampal formation. Papez suggested that the cortex of the gyrus cinguli and parahippocampal gyrus could be activated by way of association fibres from the adjacent frontal, parietal, and temporal lobes, and that the activation was perceived as joy, sorrow, anger, etc. He also pointed out that the gyrus cinguli received a large pathway from the hypothalamus via the mammillothalamic tract, anterior thalamic nucleus, and anterior thalamic radiation, and that through this pathway joy, sorrow, anger, might also be excited by changes in the environment. Thus the limbic lobe functioned, in part, as a sensory area for the perception of emotional feelings. The limbic lobe in turn could modify behaviour according to this theory by way of its own efferent cortex in the hippocampal formation.

[1] *Arch. Neurol. & Psychiat.*, XXXVIII (1937), 725.

12. The Cerebral Hemisphere III: Internal Structure

I. GENERAL FEATURES

THE GENERAL FEATURES of the internal structure of the cerebral hemisphere are revealed in a frontal section through the postcentral gyrus (figure 97B). In a central location, just lateral to the diencephalon, are the cell masses that collectively comprise the corpus striatum. At the upper and lower borders of the corpus striatum are the sections of the lateral ventricle, which is cut twice because it is a tubular structure that partially encircles the corpus striatum. The rest of the section is white matter except for the most important outer covering of grey matter, the cerebral cortex. Note that the lateral ventricle is located near the medial surface of the hemisphere and that part of its medial wall is the paper-thick choroid membrane.

II. THE WHITE MATTER

There are three kinds of fibres in the white matter of the cerebral hemisphere: association, commissural, and projection fibres. The *association fibres* are the fibres of cells in the cortex that connect cortical areas in the same hemisphere. The *commissural fibres* are the fibres that connect an area of cortex in one hemisphere with another area in the other hemisphere. The *projection fibres* connect the cortex with the corpus striatum or with the brain stem. These fibres may be either afferents or efferents of the cortex. The three are intricately interwoven to make up the white matter but the longer fibres of each category come together for part of their course to form dissectable bundles or laminae.

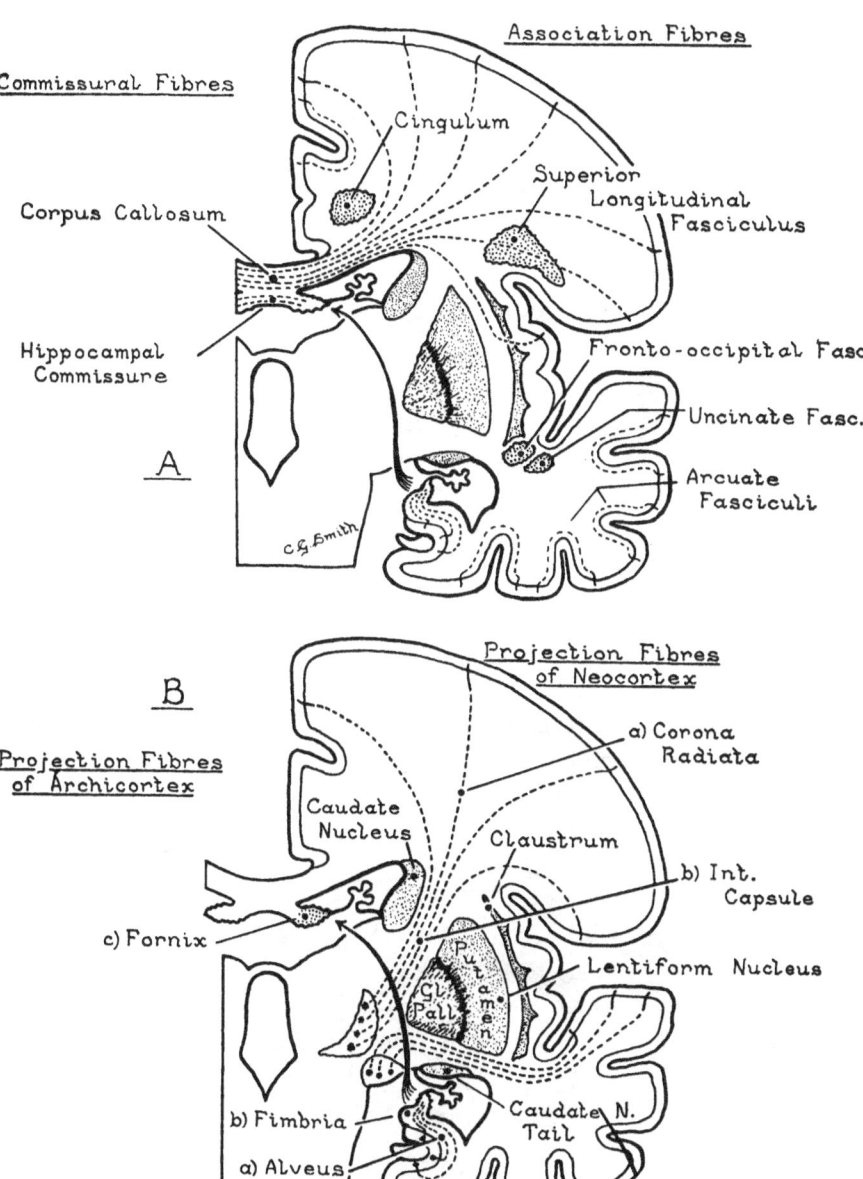

FIGURE 97. A: A frontal section of the right cerebral hemisphere to show the location of the commissural and association fibres. B: The same frontal section showing the location of the projection fibres.

A. THE BUNDLES OF ASSOCIATION FIBRES

1. THE BUNDLES OF SHORT ASSOCIATION FIBRES

Short association fibres connect adjacent gyri and are closely crowded into superficial U-shaped bands located just deep to the cortex. These are found in all parts of the hemisphere and are illustrated in the frontal section of figure 97A and in the dissection illustrated in figure 99.

2. THE BUNDLES OF LONG ASSOCIATION FIBRES

The long association fibres that connect the anterior and the posterior portions of the hemisphere come together to form three large bundles. These are located deep to the short association fibres and just deep enough to pass under the deepest of the transverse sulci, such as the central sulcus. One of these long bundles serves the cortex on the medial side of the hemisphere. The other two serve the cortex on the lateral side.

a) THE CINGULUM

This bundle of long association fibres is located in the core of the limbic lobe (figures 97A, 98). Its fibres come from and go to the cortex of the adjacent parts of the frontal, the parietal, and the occipital and temporal lobes.

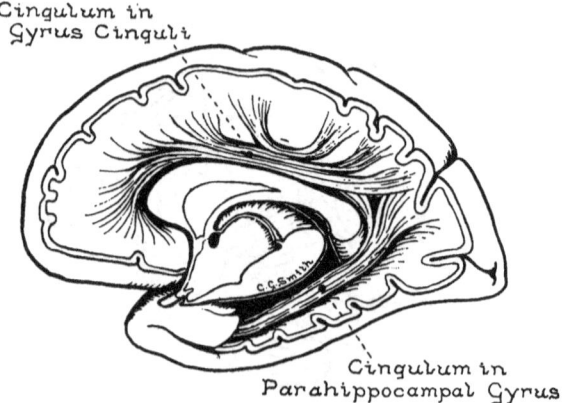

FIGURE 98. A dissection of the medial aspect of the cerebral hemisphere showing the cingulum in the core of the cingulate gyrus and the parahippocampal gyrus.

b) THE SUPERIOR LONGITUDINAL FASCICULUS

The long association fibres that course anteroposteriorly in the lateral portion of the hemisphere are crowded into an upper and a lower bundle to pass above and below the insula (figures 97A, 99). The superior bundle, the superior longitudinal fasciculus, is about 2 cm. in diameter and is most compact above the posterior border of the insula where it turns round the end of the lateral sulcus to extend into the temporal lobe. This bundle may be looked upon as an association bundle of the opercula of the insula. In the dominant hemisphere these portions of the hemisphere contain the areas of cortex essential to speech.

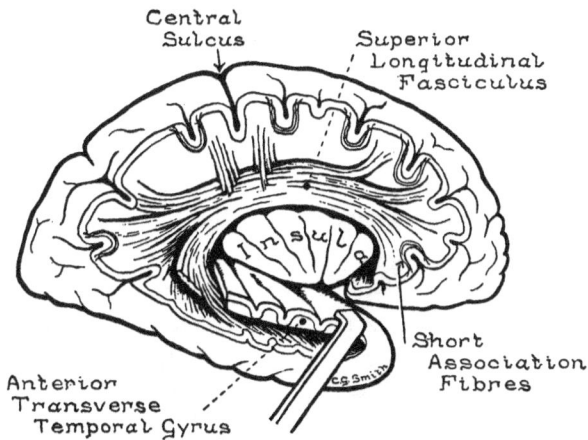

FIGURE 99. A dissection of the lateral aspect of the cerebral hemisphere showing the superior longitudinal fasciculus.

c) THE INFERIOR LONGITUDINAL FASCICULUS

The name inferior longitudinal fasciculus is used here for the compact bundle that lies deep to the lower border of the insula. Its upper fibres are the longer ones in the bundle and they can be followed forward to the frontal pole and backward to the occipital pole. Let us call them the fronto-occipital fasciculus (figures 97A, 100). The lower fibres in the inferior longitudinal fasciculus are short and form the hook-like uncinate fasciculus which connects the orbital cortex with the cortex of the temporal pole (figures 97A, 100).

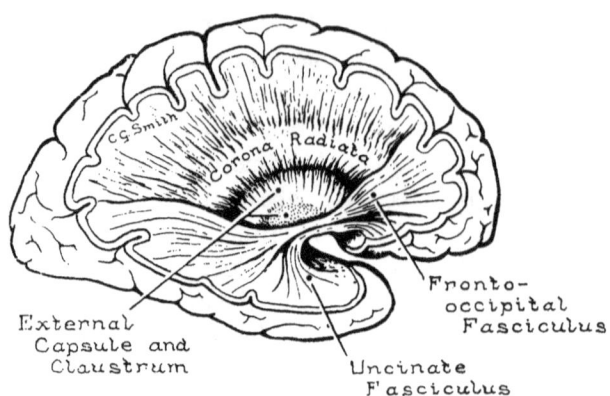

FIGURE 100. A dissection of the lateral aspect of the cerebral hemisphere showing the long association fibres coursing deep to the inferior angle of the insula.

B. THE BUNDLES OF COMMISSURAL FIBRES

The commissural fibres are grouped into three bundles: the corpus callosum, the anterior commissure, and the hippocampal commissure.

1. THE CORPUS CALLOSUM

This commissure contains all the fibres that connect the areas of neocortex in the two hemispheres except the small number in the anterior commissure. Its fibres form the upper and anterior part of the canopy-like structure that is attached to the lamina terminalis and arches back above the diencephalon to within 2 inches of the occipital pole (figures 86, 92B). The portion of the corpus callosum in the upper border of the canopy is the body of the corpus callosum. The thickened posterior free edge is the splenium, the tapered, bent back, anterior part of the corpus callosum is the rostrum, and the knee-like bend at the anterior border of the body of the corpus callosum is the genu. The genu is 1½ inches from the frontal pole.

In a frontal section the corpus callosum is seen to retain its identity within the white matter of the hemisphere (figure 97A). It crosses above and forms part of the wall of the lateral ventricle. As it proceeds laterally fibres leave its superior aspect to fan out toward the cortex. When it reaches the lateral border of the ventricle it is reduced to

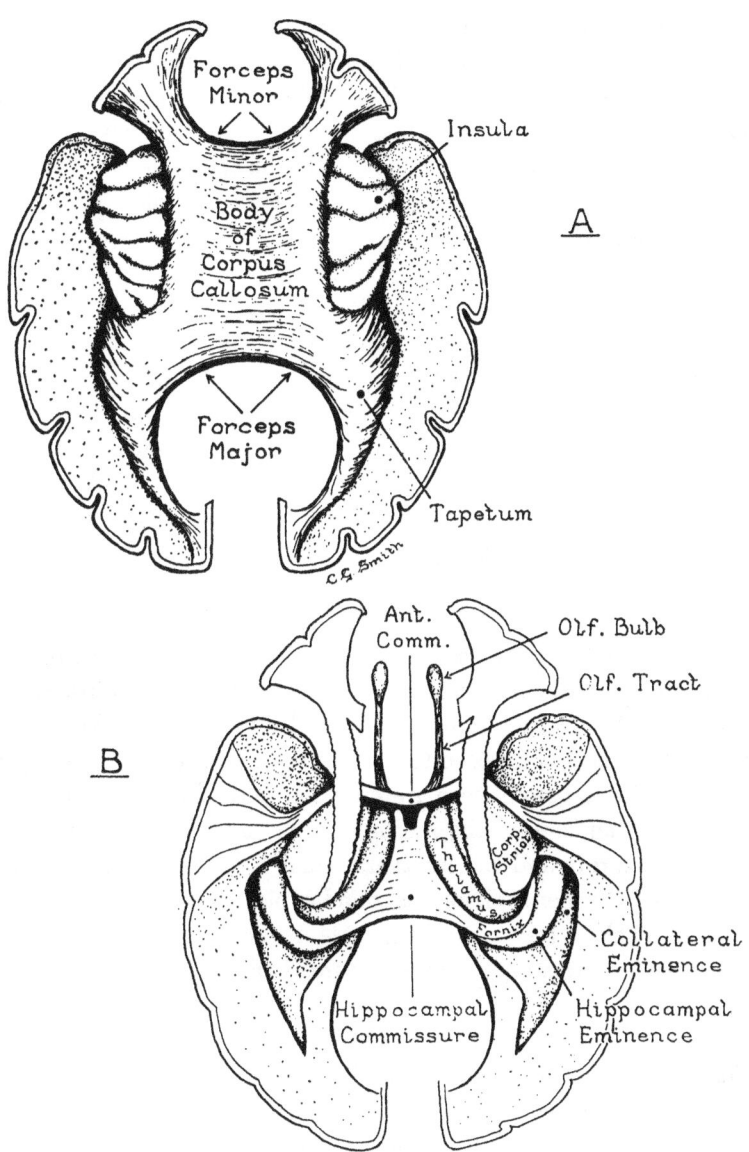

FIGURE 101. A: A dissection of the superior aspect of the cerebral hemisphere to show the corpus callosum. B: The same preparation after removing the corpus callosum, the septum pellucidum, the insula and most of the caudate and lentiform nuclei to expose the hippocampal and the anterior commissures.

a thin lamina. There it turns downwards, deep to the insula and passes through the outer part of the corpus striatum. In doing so it detaches a thin lamina-like nucleus called the claustrum, from the lens-shaped medial portion called the lentiform nucleus. Thus the fibres of the corpus callosum contribute to the shell-like layer of white matter called the external capsule, that forms the outer covering of the lentiform nucleus.

The course of the anterior and the posterior fibres of the corpus callosum can be seen in a dissection of the superior aspect of the hemispheres (figure 101A). The shorter anterior fibres course toward the frontal poles and the longer posterior fibres course toward the occipital poles to form an anterior and a posterior forceps-like structure. These are the *forceps minor* and the *forceps major*. Some of the posterior fibres course laterally behind the insula and descend in the outer wall of the posterior and inferior horns to form a dissectable lamina called the *tapetum* (carpet).

2. THE ANTERIOR COMMISSURE

This is the oldest commissure of the cerebral hemispheres. In lower animals such as fish, its fibres unite the two olfactory bulbs. In man the olfactory fibres are almost all replaced by neocortical commissural fibres that connect the neocortex in the right and left temporal poles. The fibres of this commissure are crowded into a bundle about 2 mm. in diameter that crosses the midline in the lamina terminalis, just in front of the fornix and the interventricular foramen. Traced from the midline into the hemisphere the bundle pierces the corpus striatum but passes below the internal capsule to enter and fan out in the anterior part of the temporal lobe (figure 101B).

3. THE HIPPOCAMPAL COMMISSURE

This commissure is made up of fibres that connect the archicortex of the hippocampal formation in the right hemisphere with that in the left hemisphere. In the midsagittal section it appears as a thin lamina in the inferior border of the portion of the septum pellucidum behind the interventricular foramen. Traced from the midline into the hemisphere, the fibres of this commissure turn back as part of the fornix to reach the hippocampal formation (figure 101B).

C. THE BUNDLES OF PROJECTION FIBRES

1. THE PROJECTION FIBRES OF THE ARCHICORTEX

The projection fibres of the archicortex leave the cerebral hemisphere in the bundle called the fornix. You will recall that these fibres come from cell bodies in the hippocampus and form first the alveus and then the fimbria which is the first part of the fornix. The fimbria is joined by more and more fibres as it courses along the border of the hippocampal formation (figures 94, 101B), and it increases in size accordingly. At the medial end of the hippocampal formation, just below the splenium, the fornix courses anteriorly close to the midline, at first adhering to the under side of the corpus callosum and then gradually separating from it to lie in the lower border of the septum pellucidum. As the fornix courses anteriorly next to the midline its commissural fibres leave it to cross to the other side but its projection fibres continue on toward the lamina terminalis as the column of the fornix. There, just in front of the interventricular foramen, they descend into the diencephalon.

2. THE PROJECTION FIBRES OF THE NEOCORTEX

In a frontal section (figure 97B) the projection fibres of the neocortex are seen in "pure culture" as an L-shaped band of white matter within the corpus striatum. This is the internal capsule which was described with the diencephalon. From the internal capsule the projection fibres can be followed by dissection into the diencephalon and also toward the cortex. The accepted way of demonstrating the bundles of projection fibres that fan out toward the cortex is to dissect away the lateral part of the hemisphere removing cortex of insula and its underlying white matter until the corpus striatum is exposed. When this is done the projection fibres will be seen emerging from its border (figure 100). To follow the fibres to the cortex it is necessary of course to remove all the fibres of the superior longitudinal fasciculus and inferior longitudinal fasciculus and to break off the fibres of the corpus callosum that are coursing transversely toward the lateral surface of the hemisphere. The latter break off easily except at the upper border of the corpus striatum. There they are present in such large numbers as they pass through the radiating projection fibres that they produce a dissection artefact, an artificial product. This takes the form of a crown-like ridge located above the bulging corpus striatum. The surface of this ridge is marked by radiating fibres, hence it is called the corona radiata. The radiating fibres are projection fibres and it is customary to refer to all the projection fibres within the white matter of the hemisphere as fibres of the corona radiata.

III. THE GREY MATTER

The grey matter of the hemisphere is in the cerebral cortex, the corpus striatum, and the septal region.

A. THE CEREBRAL CORTEX

The cerebral cortex has already been described. It covers all the free surface of the hemisphere except for the anterior perforated area, the septal region, and the choroid membrane. The cortex can be distinguished from other grey matter of the surface of the hemisphere because it has its cells arranged in layers. The older cortex, the archicortex, is located in or close to the hippocampal sulcus. It has three layers. The newer, that is the neocortex, forms the rest of the cortex. It has six layers. Subdivisions of the neocortex based on variations in the thickness of the cell layers and variations in cell structure conform to the functional subdivisions of the cortex. These are called cortical areas. Of these the only area recognizable in the gross is the visual sensory area. This is identified by the Line of Gennari, a white line that divides the thickness of the cortex into a superficial and a deep layer.

B. THE CORPUS STRIATUM

1. GENERAL FEATURES AND SUBDIVISIONS

This is a mass of cells in the central part of the hemisphere that occupies most of the space between the insula, laterally, and the diencephalon, medially. It is enclosed in white matter except where it comes to the surface of the hemisphere as the anterior perforated substance. Because of its location in the stalk of the hemisphere it follows that all the projection fibres uniting cerebral cortex and brain stem via the lateral aspect of the diencephalon must pass through it. Hence it is penetrated by bands of fibres which in a section appear as white streaks and give the body of grey matter its name.

In the human brain, the marginal sulcus of the insula is so deep that there is only a small interval between its floor and the lateral wall of the lateral ventricle (figure 97B). This interval is a narrow cleft-like passage through which the projection fibres must pass to reach the frontal, parietal, occipital, and temporal lobes. Hence they are crowded together and form a lamina which is somewhat like an open Japanese fan. The portion of the lamina within the corpus striatum is the internal capsule and it and the external capsule, described with the corpus callosum, are the fibre bands that incompletely separate the following four nuclei.

a) THE LENTIFORM NUCLEUS

(1) **Shape and Relationships**

The lentiform nucleus has the shape of a converging lens and is so oriented in its location lateral to the internal capsule that its principal axis passes through the centre of the insula. If it were a lens it would be in a position to focus light rays from the cortex onto the diencephalon. In a sense this is what it does with nerve impulses reaching it from the cerebral cortex. Its medial surface has three flat areas that are related to the anterior limb, the posterior limb, and the sublenticular limb of the internal capsule respectively (figures 47C, 102). The lateral surface of the nucleus is rounded and is covered by the external capsule. The anterior two-thirds of the lentiform nucleus has an inferior surface as well as a medial and a lateral. This inferior surface is directly continuous posteriorly with the infero-medial facet that is resting on the sublenticular limb of the internal capsule. The middle portion of this inferior surface reaches the inferior surface of the hemisphere and forms the portion of the surface that is outlined by the olfactory trigone, the medial and lateral olfactory striae, and the optic tract. This area has openings for fine vessels that supply the corpus striatum and it is called accordingly the anterior perforated area. Behind the anterior perforated area the inferior surface of the lentiform nucleus rests on and is joined to another nucleus of the corpus striatum, the amygdaloid nucleus. Behind this connection with the amygdaloid nucleus the lentiform nucleus rests on the sublenticular limb of the internal capsule. In front of the anterior perforated area the inferior surface of the

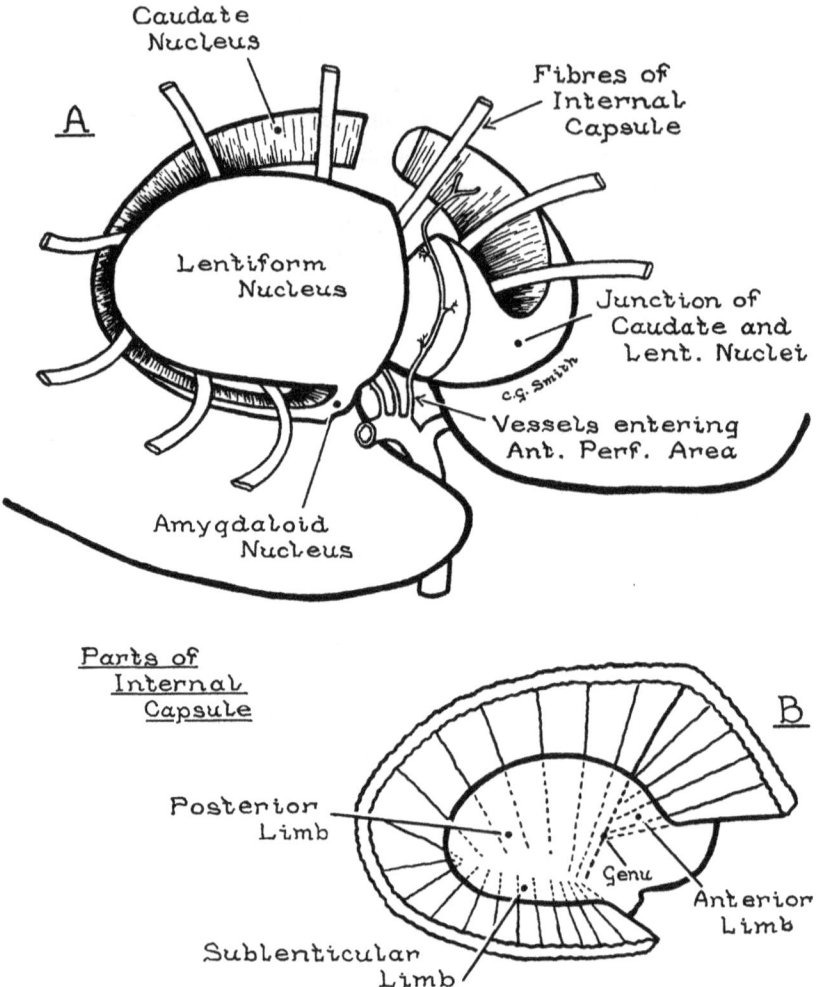

FIGURE 102. A: The lateral aspect of the lentiform, the caudate and the amygdaloid nuclei. A segment has been removed from the lentiform nucleus in the region of the anterior perforated area. B: The above preparation before removing the fibres of the internal capsule and the segment of the lentiform nucleus.

lentiform nucleus is covered by white matter and here the lentiform nucleus is joined on its medial side to a third nucleus of the corpus striatum, the caudate nucleus. This connection with the caudate is just below the anterior limb of the internal capsule (figure 102).

(2) The Subdivisions of the Lentiform Nucleus

The lentiform nucleus is divided into an upper, larger part, the lentiform nucleus proper, which is immediately lateral to the internal capsule and a lower, smaller, anterior part, the innominate substance, which is immediately lateral to the hypothalamus and the septal area of the hemisphere. The horizontal plane between the upper and the lower parts of the lentiform nucleus is the plane of the anterior commissure which passes transversely through the corpus striatum to reach the midline.

The superior part, the lentiform nucleus proper, is divided into a medial and a lateral portion. The lateral portion is the *putamen* (a paring). It contains medium-sized cells and relatively few myelinated fibres. The medial portion of the lentiform nucleus proper is the *globus pallidus* (pale body). It contains large cells and many myelinated fibres, hence it is paler than the putamen in a freshly cut brain (figure 97B). The putamen receives the afferent fibres that come to the lentiform nucleus, and the large cells of the globus pallidus give rise to most of the efferent fibres of the lentiform nucleus.

The inferior part of the lentiform nucleus, the so-called innominate substance, extends to the inferior surface of the hemisphere where it forms the substance of the anterior perforated area. Its cells are small, in clusters, and irregularly arranged.

b) THE CAUDATE NUCLEUS

The caudate nucleus is a somewhat ribbon-like semicircular band of grey matter that becomes narrow toward one end, like a tail, hence its name (figure 102A). It is helpful to consider it as a detached marginal portion of the putamen (figure 97B). This thick ribbon of grey matter has a ventricular surface that is rounded and a capsular surface that is flat. The anterior end of the caudate nucleus is very wide and thick. The part in front of the interventricular foramen is called the head. Posteriorly, the caudate nucleus grows narrower and thinner, and the end portion in the temporal lobe is called its tail. The end of the tail of the caudate enlarges to form the amygdaloid nucleus which is partly in the wall of the ventricle and partly anterior to it. The head of the caudate located in the frontal extremity of the lateral ventricle extends inferiorly beyond the lateral ventricle to join the putamen of the lentiform nucleus below the anterior limb of the internal capsule.

c) THE AMYGDALOID NUCLEUS

The amygdaloid nucleus (like an almond) is a flattened, ovoid mass about the size and shape of an almond. It lies in the upper wall of the temporal extremity of the lateral ventricle and is connected with the tail of the caudate nucleus. Superiorly and posteriorly it is separated from the lentiform nucleus by the sublenticular part of the internal capsule but anteriorly it is directly continuous with the innominate substance and with the putamen. Medially and anteriorly it comes to the surface of the uncus.

d) THE CLAUSTRUM

The claustrum (a barrier) is a thin, shell-like layer of grey matter that is separated from the outer surface of the putamen by the external capsule. Its inferior border is thick and it tapers to a thin edge superiorly. It is coextensive with the insular cortex and is separated from it by a very thin layer of white matter.

2. THE FUNCTION AND CONNECTIONS OF THE CORPUS STRIATUM

The corpus striatum is a portion of the motor system. Part of it controls striated muscles and part of it controls smooth muscle and glands. The caudate and the putamen and the globus pallidus are relay stations on pathways to striated muscles. The amygdaloid nucleus, the claustrum, and the innominate substance have visceral functions.

Impulses are brought to the caudate and putamen from the motor cortex and from the medial nuclei of the thalamus (the latter via the inferior thalamic peduncle). These impulses are relayed to the globus pallidus and thence via the ansa and fasciculus lenticularis to the subthalamus. From the subthalamus, pathways with relay stations in the reticular formation cross the midline and descend to the motor nuclei. Thus the corpus striatum influences motor activity directly. In addition to this direct influence, the corpus striatum can influence it indirectly through a feedback pathway that extends from the subthalamus to the ventral lateral nucleus and thence to the motor cortex.

The caudate and the putamen are considered by physiologists to form a unit; they call it the striatum. The globus pallidus serves this unit as a dispatching centre and they call it the pallidum.

The amygdaloid nucleus receives some of the fibres of the lateral olfactory tract. It also receives pathways from the septal area and the hypothalamus via the innominate substance, that is, the anterior perforated region. The data brought by these pathways come chiefly from visceral sources. Of the efferent pathways of the amygdaloid nucleus some course medially through the innominate substance to reach the septal area and the hypothalamus. If these fibres and the afferents of the amygdaloid nucleus coursing through the innominate substance are superficial they form a pale strip along the posterior border of the anterior perforated area called the **diagonal band.** Other efferent fibres of the amygdaloid nucleus reach the hypothalamus by a very roundabout route. They come together and form a thread-like band of myelinated fibres that follows the medial border of the caudate in the wall of the inferior horn and the body of the lateral ventricle as far as the interventricular foramen. This band is the stria **terminalis.** It joins the fornix just in front of the interventricular foramen.

IV. THE LATERAL VENTRICLE

A. FORM AND SUBDIVISIONS

The lateral ventricle is a curved tubular space that extends anteroposteriorly in the long axis of the frontal, parietal, and temporal lobes, and has a recess directed backwards into the occipital lobe (figure 103). Its form is explained by its development. You will recall that the initially ovoid cavity of the embryonic cerebral hemisphere is encroached on from below by a thickening of the middle part of its floor that develops into the corpus striatum. This mass bulging upwards fills all but the anterior, superior, and posterior portions of the cavity and thus converts it into a curved tubular space that partially encircles the corpus striatum. Later as the walls of the occipital portion of the hemisphere grow thicker and encroach on the cavity, a pouch-like recess is left extending back toward the occipital pole.

The lateral ventricle has four subdivisions, (1) an anterior horn

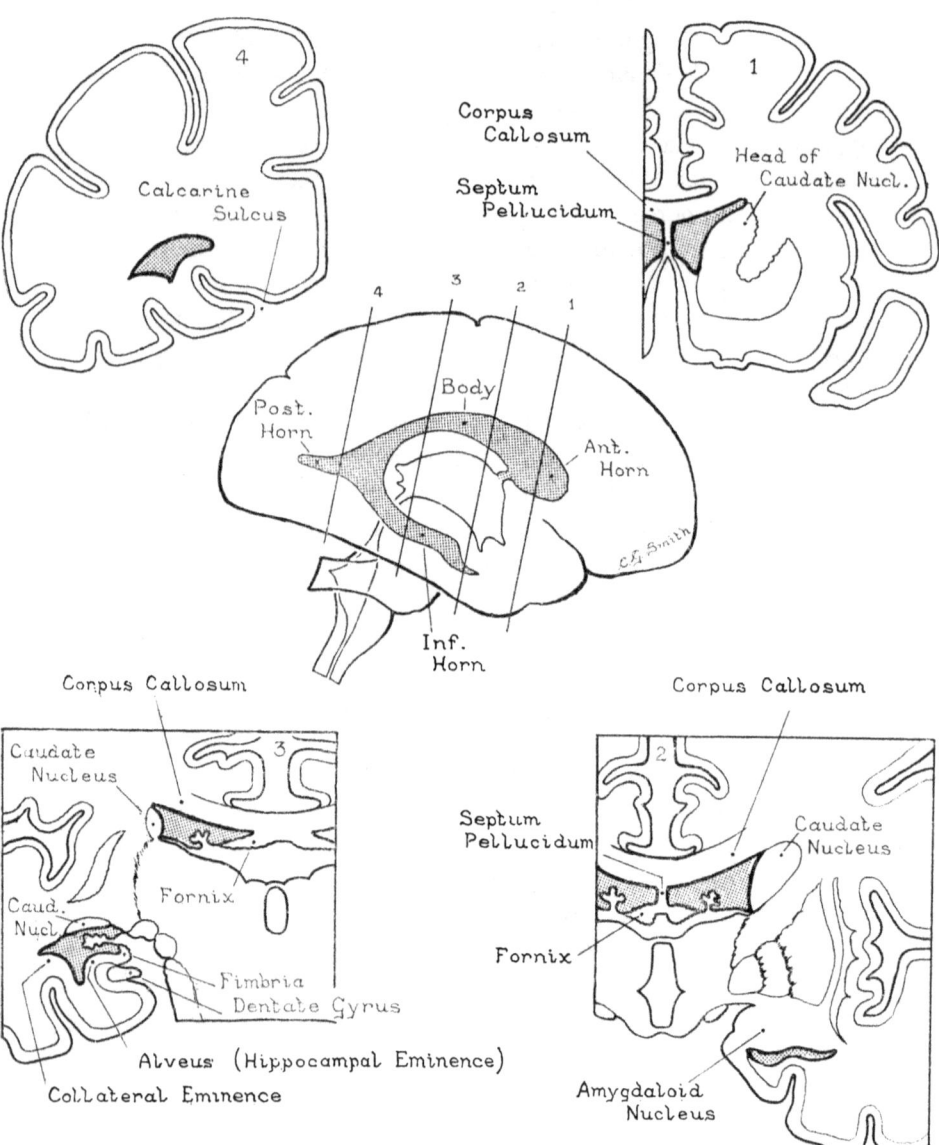

FIGURE 103. Four serial frontal sections of the cerebral hemisphere at the sites indicated in the central diagram to show the structures in the wall of the lateral ventricle.

located in the frontal lobe in front of the interventricular foramen, (2) a body which extends backwards from the interventricular foramen to the middle of the parietal lobe where it forks into (3) the posterior horn that extends back into the occipital lobe and (4) the inferior horn that extends inferiorly and then forward into the temporal lobe.

B. STRUCTURES IN THE WALL OF THE LATERAL VENTRICLE

The structures in the wall of the lateral ventricle are shown in the frontal sections illustrated in figure 103. These structures can be seen to advantage in a dissection of the brain if the ventricle is opened from end to end through the portion of its wall formed by the corpus callosum. Such a cut should begin near the genu of the corpus callosum about ¼ inch from the midline and then be extended forward and back looking into the ventricle all the while to avoid cutting into the caudate nucleus.

In such a preparation the structures of the wall of the anterior horn, body, and inferior horn can be seen to form a series of parallel bands. In the body of the ventricle we can easily identify the caudate nucleus and follow it forwards to the tip of the anterior horn and backwards to the tip of the inferior horn. It tapers from front to back and ends at the tip of the inferior horn in an enlargement which is the amygdaloid nucleus. Medial to the caudate nucleus in the body of the ventricle is the choroid membrane with its choroid plexus. This membrane is paper-thick and ribbon-like and has one edge attached to the groove between the diencephalon and the caudate nucleus. The other edge of the choroid membrane is attached to the border of the fornix. This membrane extends along the length of the caudate from the level of the interventricular foramen to a point near the end of the inferior horn. The choroid plexus has a swelling called the glomus at the junction of the posterior and the inferior horns (figure 104). This may contain large cysts and calcified nodules and in that event it is a landmark for the radiologist.

Medial to the choroid membrane in the body of the ventricle is the fornix. This band of fibres is coextensive with the choroid membrane and tapers to a point near the tip of the inferior horn. In the inferior horn the fornix is called the fimbria because it is the fringe-like border of the hippocampal eminence. This eminence is a ridge produced by the hippocampal sulcus. The ridge has a rounded end at the tip of

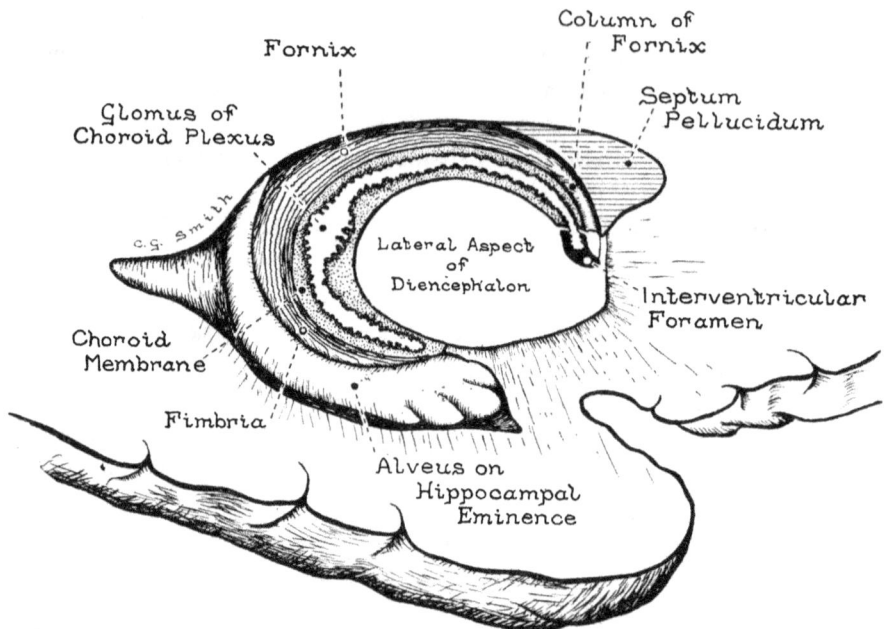

FIGURE 104. The structures that can be seen in the medial portion of the lateral ventricle by removing the caudate nucleus and the tapetum.

the inferior horn and tapers to a point where the inferior horn joins the body of the lateral ventricle.

Medial to the fornix in the anterior part of the body of the ventricle we find the septum pellucidum. This membrane is in the midsagittal plane. It is like a window-pane set in a frame formed by the corpus callosum above and the fornix plus the rostrum of the corpus callosum below. The upper and lower portions of the posterior part of the frame come together just in front of the splenium and here the septum pellucidum tapers to a point.

In the inferior horn, lateral and parallel to the hippocampal eminence, is the collateral eminence. This is a ridge produced by the collateral sulcus.

The posterior horn has thick walls of white matter all around. However, the calcarine sulcus is deep enough to produce an antero-posterior ridge here called the *calcar avis* (cock's spur). There may be another ridge in the posterior horn above the calcar avis; if present it is produced by the fibres of the forceps major.

13. The Autonomic Nervous System

I. DEFINITION AND CHARACTERISTICS

THE AUTONOMIC nervous system is a subdivision of the *motor* portion of the nervous system. It carries impulses to a special group of effectors, smooth muscles, heart muscle, and glands. These effectors

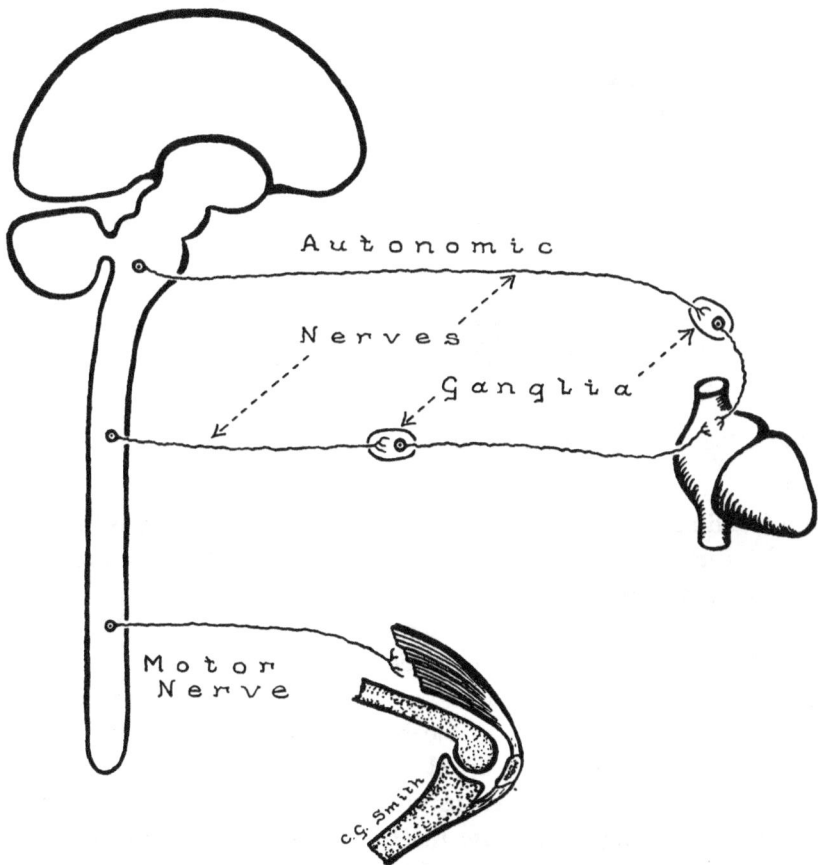

FIGURE 105. The difference between motor nerves to striated muscles and the autonomic nerves.

208   BASIC NEUROANATOMY

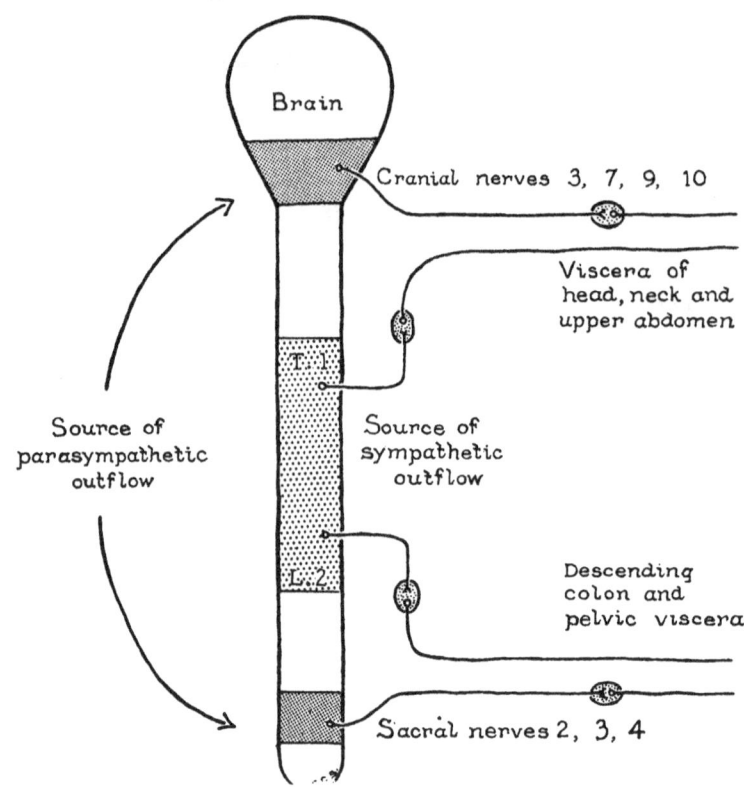

FIGURE 106. The sympathetic and the parasympathetic portions of the autonomic nervous system.

are found in viscera and also in non-visceral structures such as the skin.

The autonomic nervous system has two distinguishing features. (1) Its motor nerves to the viscera are paired, one acting as an excitor and one as an inhibitor. (2) Each of the "motor nerves" of this system is a chain of two nerve cells with the cell body of the distal nerve cell in a peripheral nerve ganglion. To illustrate these two characteristics let us consider the nerves to the heart (figure 105). The heart is a muscle and therefore it should be activated like the quadriceps femoris by one motor nerve. Instead, its activity is influenced by two nerves, one acting to increase the frequency of its contractions, the other to decrease the frequency. Each of these two nerves has its own cell station (relay station) in the peripheral nervous system. These are autonomic ganglia and will be described later.

II. THE SYMPATHETIC AND THE PARASYMPATHETIC DIVISIONS OF THE AUTONOMIC NERVOUS SYSTEM

The two motor nerves to each of the viscera come from widely separated parts of the central nervous system. One of the pair comes from a spinal cord segment between Th.1 and L.2. The other comes from either the brain or the sacral part of the cord. The nerves emanating from the thoracic part of the cord and from the first two lumbar segments comprise the sympathetic division of the autonomic nervous system (figure 106). The nerves emanating from the brain and sacral part of the cord comprise the parasympathetic (*para*, on either side) portion of the autonomic nervous system.

A. THE SYMPATHETIC DIVISION

The fibres of the sympathetic nerves leave the spinal cord in the ventral roots of nerves Th.1 to L.2. They course distally into the anterior ramus to a point just lateral to the vertebral column. There

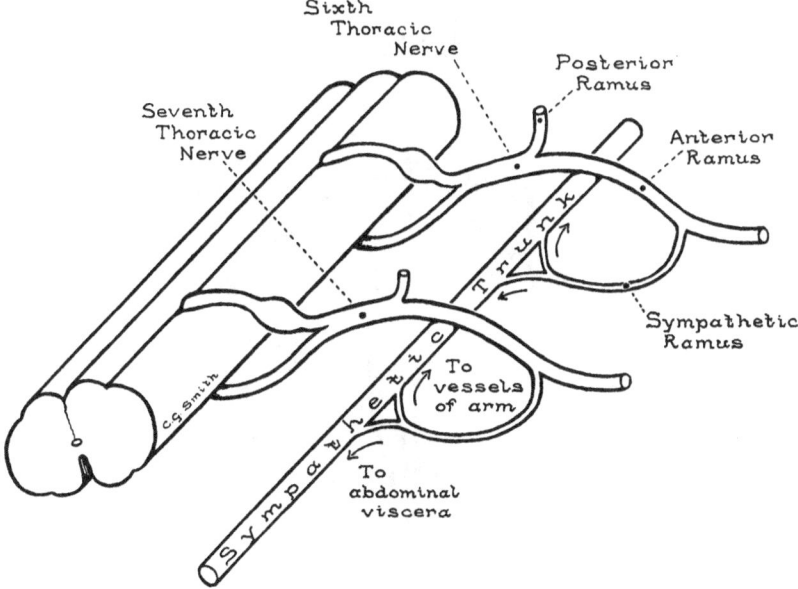

FIGURE 107. The formation of the longitudinal paravertebral nerve called the sympathetic trunk.

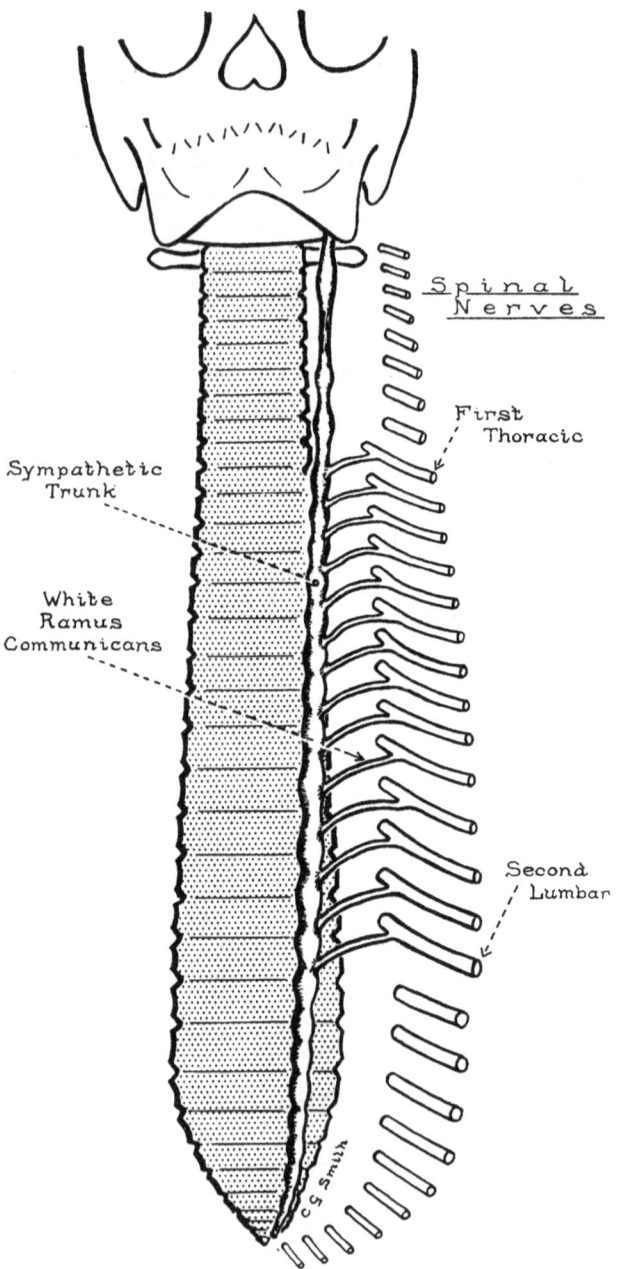

FIGURE 108. The sympathetic trunk and its afferent branches, the white rami communicantes.

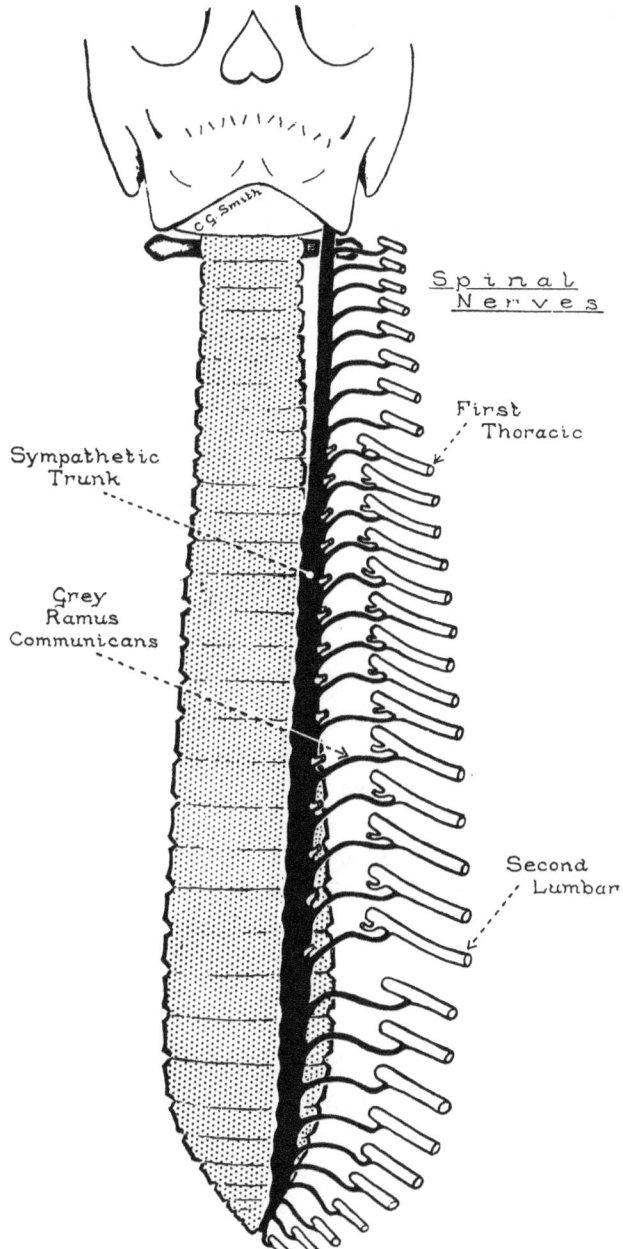

FIGURE 109. The sympathetic trunk and its efferent branches, the grey rami communicantes.

they leave the anterior ramus and course medially in a branch called the white ramus communicans (figure 107). The name of this branch is fitting because the fibres are myelinated and it connects a spinal nerve with the sympathetic trunk to be described below.

The white rami pass medially to the side of the vertebral column and there some of the nerve fibres of each ramus turn at right angles to ascend a certain distance toward the head, while others turn to descend a certain distance toward the coccyx. The longitudinally running fibres of the fourteen nerves come together as they ascend or descend to form a cable-like bundle called the sympathetic nerve trunk, which lies alongside the vertebral bodies. It extends all the way from the base of the skull to the tip of the coccyx (figure 108). It is through this nerve trunk that the sympathetic nerves are distributed to the head and neck, and to the pelvis and lower limb. The sympathetic trunk, therefore, is a device to ensure the distribution of impulses to smooth muscles and glands throughout the length of the body. Let us turn now to a description of the nerves that leave the sympathetic trunk to distribute these impulses.

1. THE SEGMENTAL NERVES

The sympathetic nerve fibres for each segment leave the lateral side of the sympathetic trunk in a thread-like nerve called the grey ramus communicans. This nerve is grey because the nerve fibres are unmyelinated; they are coming from cells located within the trunk as will be described later. One of these nerves enters the anterior ramus of each of the spinal nerves (figure 109). By entering the spinal nerve the sympathetic nerve fibres are able to reach all parts of the segment of the body. To reach the posterior part of the body some fibres must pass toward the cord for a short distance to find their way into the posterior ramus.

By comparing the drawings showing the grey and the white rami of the sympathetic trunk, in figures 108 and 109, it will be evident that the anterior rami of the spinal nerves Th.1 and L.2 have both a white and a grey ramus and that all the other nerves have only a grey ramus. The two rami of the second lumbar nerve and the single grey ramus of the third lumbar nerve are illustrated in figure 110. Cutting a grey ramus would be one way of cutting all the sympathetic fibres to a body segment. This is sometimes necessary to alleviate an

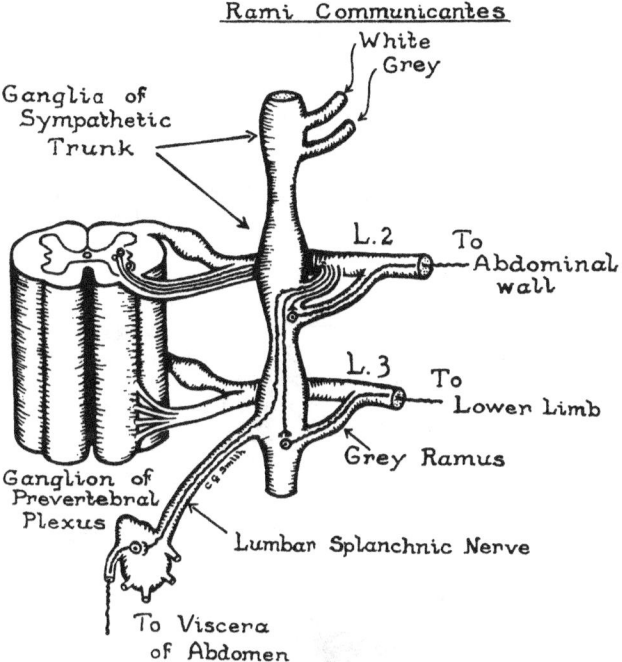

FIGURE 110. Three sympathetic pathways. (1) To the sweat glands of a segment supplied by a nerve that has a white ramus communicans (L. 2). (2) To the sweat glands of a segment supplied by a nerve that lacks a white ramus communicans (L. 3). (3) A pathway to the abdominal viscera.

abnormal condition that is brought about by an overactive sympathetic nerve. An example of such an abnormal condition is impaired blood supply due to excessive vasoconstriction.

2. THE VISCERAL NERVES

The nerves leaving the sympathetic trunk for the viscera run from the medial side of the trunk (figure 111). They course toward the midline to reach the viscera of the neck, the thorax, and the abdomen. These are called the carotid, cardiac, greater splanchnic, lesser splanchnic, lowest splanchnic, and lumbar splanchnic nerves. Some unnamed nerves leave the sacral part of the sympathetic trunk to go to the viscera of the pelvis. All these visceral nerves pass toward the front of the vertebral column where they form networks called prevertebral plexuses. The prevertebral plexus in the neck is called the carotid plexus. In the thorax there is a cardiac, a pulmonary, and an

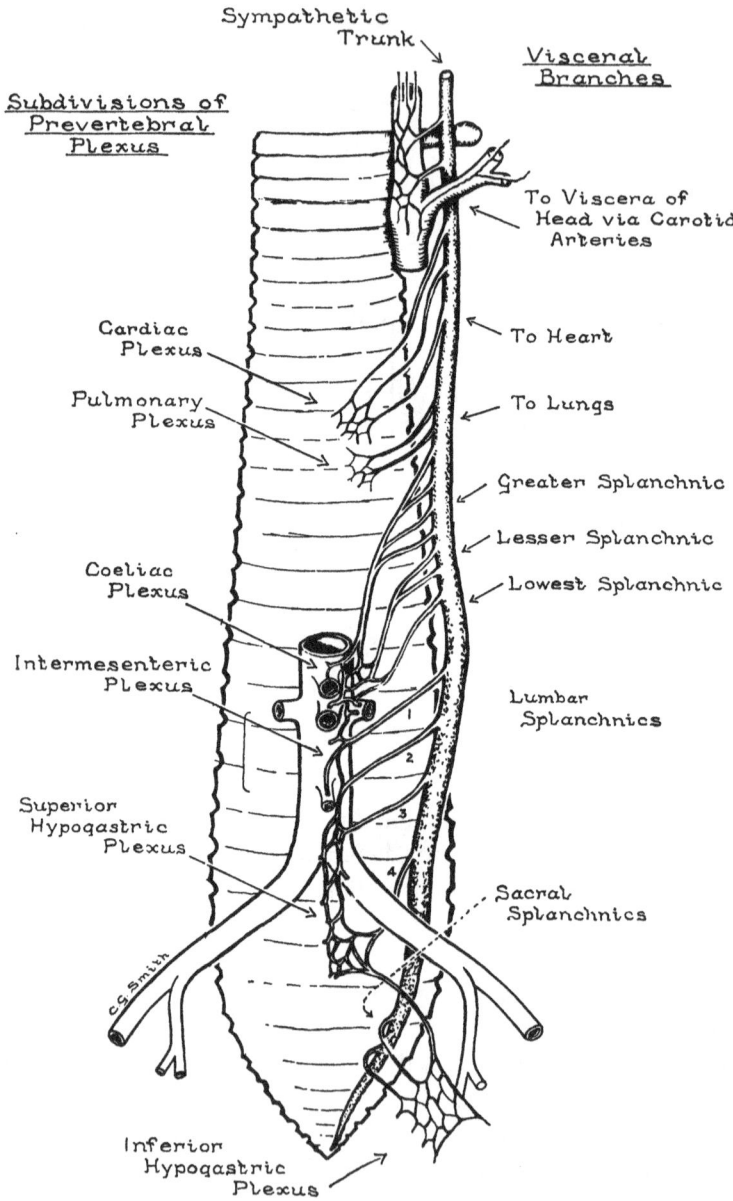

FIGURE 111. The visceral branches of the sympathetic trunk and the parts of the prevertebral plexus which they help to form.

oesophageal subdivision. In the abdomen the prevertebral plexus extends from the level of the coeliac artery to the second piece of the sacrum where it forks into a right and a left portion. Each of these pelvic portions of the plexus forms a quadrilateral sheet that is interposed between the pelvic viscera and the vessels on the pelvic wall. The upper, median part of the plexus in the abdomen is divided for descriptive purposes into a coeliac portion, located around the root of the coeliac artery, an intermesenteric portion, located between the level of the origin of the superior mesenteric artery and that of the inferior mesenteric artery, and a caudal part called the superior hypogastric plexus. The pelvic plexus—divided into a right and a left portion—is the inferior hypogastric plexus. The nerve fibres leave each plexus and course along one of the visceral arteries to reach the organ they supply.

3. THE GANGLIA OF THE SYMPATHETIC NERVES

It was pointed out earlier that the motor nerves of the autonomic nervous system are peculiar in that they are a chain of two nerve cells. Hence somewhere along each sympathetic nerve there must be a collection of cell bodies—a relay station. Each of these cell masses is a sympathetic ganglion and it may be located in one of two places, either in the sympathetic trunk or in a prevertebral plexus.

a) THE GANGLIA OF THE SYMPATHETIC TRUNK

The ganglia of the sympathetic trunk form a longitudinal series of swellings, one in front of each spinal nerve. These develop as discrete ganglia, one for each segment but later the cells of adjacent ganglia may move together and form one mass. Thus it happens that the ganglia of segments C.1, C.2, C.3, and C.4 form one long fusiform superior cervical ganglion; C.5 and C.6 form a single nodular middle cervical ganglion and C.7 and C.8 form the inferior cervical ganglion. Usually the first thoracic ganglion and the inferior cervical unite to form a composite mass called the stellate ganglion.

In the thoracic region there are usually ten ganglia, in the lumbar region four, and in the sacral region three or four. The coccygeal ganglia of the right and left trunks may come together to form an unpaired ganglion impar.

b) THE GANGLIA OF THE PREVERTEBRAL PLEXUSES

The ganglia of the prevertebral plexuses are found only below the level of the diaphragm. They are irregularly distributed within the abdominal plexuses but they tend to cluster around the origin of the coeliac, the superior mesenteric, and the inferior mesenteric arteries. In the pelvis there are large ganglia in the inferior hypogastric plexus of each side.

4. THE LOCATION OF THE RELAY STATIONS OF EACH OF THE SEGMENTAL AND VISCERAL SYMPATHETIC NERVES

Each of the *segmental nerves* has its relay station in a ganglion in the sympathetic trunk at the level of the segment it supplies. Each of the *visceral nerves* to the structures in the head, neck, and thorax also has its relay station in a ganglion in the sympathetic trunk. The nerves supplying structures in the *head* have their relay station in the superior cervical ganglion. The visceral nerves to the *heart* have a relay station in one of the cervical ganglia or in one of the upper four thoracic ganglia. Cardiac nerves may have some relay stations in the cervical ganglia because the tubular heart develops in part in the cervical region. The visceral nerves to the *lungs* have their relay stations in the upper four thoracic ganglia. The visceral nerves to the *abdomen* and *pelvis* have their relay stations in the ganglia of the prevertebral plexus at a level corresponding to the position of the viscus supplied. For example, those to the *stomach* have relay stations in ganglia in the coeliac plexus whereas those to the *bladder* and *rectum* have relay stations in the inferior hypogastric plexus.

5. THE AFFERENT FIBRES IN THE SYMPATHETIC NERVES

The nerves of the sympathetic nervous system are motor but they also contain some sensory fibres. These sensory nerve fibres come from sense organs in the viscera. They course toward the spinal cord via the splanchnic nerves as far as the sympathetic trunk and then course upwards or downwards as the case may be to reach the level of the spinal nerve which is to convey them to the central nervous system. There they leave the sympathetic trunk in a white ramus to enter the spinal nerve, course toward the cord and enter it via a dorsal root. The cell bodies of these visceral sensory fibres are in the dorsal root ganglion.

A knowledge of the segmental origin of the sensory nerve fibres of each of the viscera is of value in interpreting pain referred to the body wall from a diseased organ. Pain emanating from the heart is felt in the chest in the distribution of nerves Th.1 to Th.5. Oesophageal pain is referred to the distribution of nerves Th.5 and Th.6. Stomach pain is referred to segments Th.7, Th.8, and Th.9, chiefly in the epigastrium. Pain in the small intestine is referred to body segments Th.9 and Th.10. Appendix pain is referred to Th.10 in the umbilical region. Pain emanating from the ascending and the transverse colon is referred to segments Th.10, Th.11, and Th.12. The descending colon and part of the rectum receive pain fibres from segments L.1 and L.2, the kidney receives its fibres from segments Th.12, L.1, and L.2; the ureter, from L.1 and L.2.

The pain fibres from the bladder follow the parasympathetic nerves of segments S.2, S.3, and S.4 to the cord. The body of the uterus is supplied with pain fibres from Th.11, Th.12, L.1, and L.2, through the sympathetic nerves, but the cervix receives its pain fibres from the sacral segments 2, 3, and 4, via the parasympathetic nerves.

B. THE PARASYMPATHETIC DIVISION OF THE AUTONOMIC NERVOUS SYSTEM

The nerve fibres of the parasympathetic division leave the central nervous system in cranial nerves 3, 7, 9, and 10, and in sacral nerves 2, 3, and 4. These parasympathetic nerve fibres differ in their distribution from those of the sympathetic nervous system in that they go almost exclusively to the viscera. They lack a segmental distribution, hence there is no need for a parasympathetic trunk.

Each of the parasympathetic nerves leaves its parent cranial nerve 3, 7, 9, or 10 and its parent spinal nerve, that is, sacral nerve 2, 3, or 4, and proceeds on its own to the organ it is to supply. Some cranial nerves give off more than one parasympathetic nerve to supply more than one organ. The autonomic ganglia, that is, parasympathetic ganglia, on each of these nerves are close to or are embedded in the organ supplied. To describe the parasympathetic nervous system more fully it will be necessary to deal with it piece-meal, that is, nerve by nerve.

1. THE PARASYMPATHETIC BRANCH OF CRANIAL NERVE 3

This arises from the inferior division of the third nerve at the apex of the orbit and supplies the smooth muscles in the eye. The relay station is in the ciliary ganglion. This is about the size of a pin's head and is flattened against the optic nerve by the lateral rectus. The postganglionic fibres (fibres of cells in the ganglion) leave the ganglion as a tuft of thread-like nerves called the short ciliary nerves. They pierce the sclera around the optic nerve and course forward deep to the sclera to reach the ciliary muscle (accommodation) and also the sphincter of the pupil (light reflex).

2. THE PARASYMPATHETIC BRANCHES OF CRANIAL NERVE 7

a) THE GREATER SUPERFICIAL PETROSAL NERVE

This supplies the glands of the nose, orbit, and palate. It leaves the seventh nerve in the petrous bone and travelling medially it gradually rises through the floor of the middle cranial fossa. It is so named because it is the larger of two nerves on the superior surface of the petrous bone. The lesser superficial petrosal is a parasympathetic nerve from cranial nerve 9.

The greater superficial petrosal nerve leaves the middle cranial fossa by descending in the foramen lacerum until it reaches the opening of the pterygoid canal. Here it is joined by its fellow sympathetic nerve called the deep petrosal nerve which has come up from the superior cervical ganglion alongside the internal carotid artery. The composite nerve called the nerve of the pterygoid canal enters its canal and is conducted to the pterygopalatine fossa where it ends in its relay station, the sphenopalatine ganglion. The postganglionic nerve fibres join the maxillary nerve and are distributed with its branches to the nose, the orbit, and the palate.

b) THE CHORDA TYMPANI

This nerve, the nerve to the submandibular and the sublingual salivary glands, leaves the seventh nerve in the petrous bone just above the stylomastoid foramen. It courses medially across the eardrum and thus acquires its name. At the anterior (medial) border of the ear-drum it passes inferiorly to emerge from the skull through the petrotympanic fissure and then passes forward as it descends to join the lingual nerve. It follows the lingual nerve to the side of the tongue where it leaves it, to end in its relay station, the submandibular ganglion. The postganglionic fibres are grouped into filaments that are distributed to the closely associated submandibular and sublingual glands.

3. THE PARASYMPATHETIC BRANCH OF CRANIAL NERVE 9

This is the tympanic branch of the ninth nerve. It leaves the ninth nerve as it passes through the jugular foramen, penetrates the base of the skull to reach the medial wall of the middle ear (tympanum), and continues on through the roof of the middle ear to course alongside the greater superficial nerve. Here it is called the lesser superficial petrosal nerve. It travels medially to the mandibular foramen, where it descends to its relay station, the otic ganglion located just below the foramen at the medial side of the mandibular nerve. The postganglionic fibres join the mandibular nerve and are distributed with its auriculotemporal branch to the parotid gland.

4. THE PARASYMPATHETIC BRANCHES OF THE VAGUS NERVE

Most of the fibres of the vagus are parasympathetic. They leave the nerve in branches that come off in the neck, thorax, and abdomen.

a) THE CARDIAC NERVES

These leave the vagus partly in the neck and partly in the thorax. They course to the base of the heart, where they enter the prevertebral plexus formed by the cardiac branches of the sympathetic trunk, and accompany the sympathetic fibres to the heart. The relay stations, that is, cell clusters for the parasympathetic fibres, are either in the plexus or on the surface of the heart.

b) THE PULMONARY NERVES

The parasympathetic nerves to the lungs leave the vagus as it passes behind the root of the lung. They help form the pulmonary plexus which was described earlier as a sympathetic prevertebral plexus. The relay stations of these parasympathetic fibres are clusters of cells in this plexus or in the substance of the lung. The postganglionic fibres follow the bronchial tree to the smooth muscles and glands in its wall.

c) THE OESOPHAGEAL BRANCHES

These leave the vagus nerve as it courses alongside the oesophagus. The relay stations are in the wall of the oesophagus.

d) THE BRANCHES TO THE ABDOMINAL VISCERA

Some branches to the stomach reach its anterior surface from the plexus formed by the right and left vagus nerves on the oesophagus. These supply the stomach and the first part of the duodenum and the liver as well. Others follow the left gastric artery to the upper end of the prevertebral plexus where the parasympathetic fibres are distributed with their fellow sympathetic fibres. Their destinations are the parts of the alimentary tract as far as the left colic flexure, liver, pancreas, and kidney. The relay stations of these parasympathetic nerves are embedded in the viscera. In the alimentary tract the cell bodies of these ganglia are located between the layers of circular and longitudinal muscle, and also in the submucosa.

e) THE PARASYMPATHETIC BRANCHES OF SACRAL NERVES 2, 3, 4

These parasympathetic nerves leave the anterior rami of sacral nerves 2, 3, and 4 to supply pelvic viscera and are called the pelvic splanchnics. They pass forward on the lateral side of the rectum and bladder into the pelvic part of the sympathetic prevertebral plexus. The parasympathetic fibres are distributed to the pelvic viscera and the descending colon along with the sympathetic fibres. To reach the descending colon they ascend in the prevertebral plexus to the level of origin of the inferior mesenteric artery which conducts them to their destination.

5. THE AFFERENT FIBRES IN THE PARASYMPATHETIC NERVES

The parasympathetic nerves are visceral efferent, that is, motor to smooth muscle and glands. Nevertheless the viscera they supply contain sense organs and sensory fibres course back to the central nervous system in these nerves. Most of the sensory fibres are a variety of special sensory fibre; they come from stretch receptors in the lung, chemoreceptors in taste buds, and chemoreceptors in the carotid body.

All the sensory fibres, including some pain fibres, have their cell bodies in the sensory ganglion of the nerve supplying the organ with parasympathetic fibres, that is, in the sensory ganglia of cranial nerves 7, 9, and 10 and in the dorsal root ganglia of sacral nerves 2, 3, and 4.

14. The Blood Supply of The Brain and Spinal Cord

I. GENERAL FEATURES

BLOOD IS BROUGHT to each side of the brain by two vessels, the vertebral and the internal carotid arteries (figure 112). These two arteries enter the cranial cavity at opposite ends of the brain stem, course toward each other on its ventral surface, and anastomose to form an arterial trunk. A branch of this trunk descends onto the ventral surface of the cord to form the upper end of an arterial trunk for the spinal cord.

The longitudinal arterial trunks of the brain, a right and a left, anastomose with each other in front of the pons to form a median vessel and in front of the lamina terminalis where they are connected by a short, transverse branch. Side branches of these trunks course medially and laterally on the surface of the brain and these, particularly the ones that extend onto the cerebral hemisphere and onto the cerebellum, anastomose freely. Thus the brain is enclosed in a network of vessels through which arterial blood may reach any part of its surface.

The nutrient vessels of the brain are branches of the vascular tunic. They are fine, thread-like branches that penetrate the brain at right angles to its surface. Each vessel is the core of a cylindrical or cone-shaped portion of nervous tissue which it nourishes. Its capillaries anastomose with those of adjacent nutrient arteries but this anastomosis is not free enough to permit one vessel to supply the capillary bed of its neighbour.

II. THE PARTS OF THE ARTERIAL TRUNK OF THE BRAIN

The **vertebral artery** enters the skull through the foramen magnum at the side of the junction of cord and medulla and ascends to join its

BLOOD SUPPLY OF BRAIN AND SPINAL CORD 223

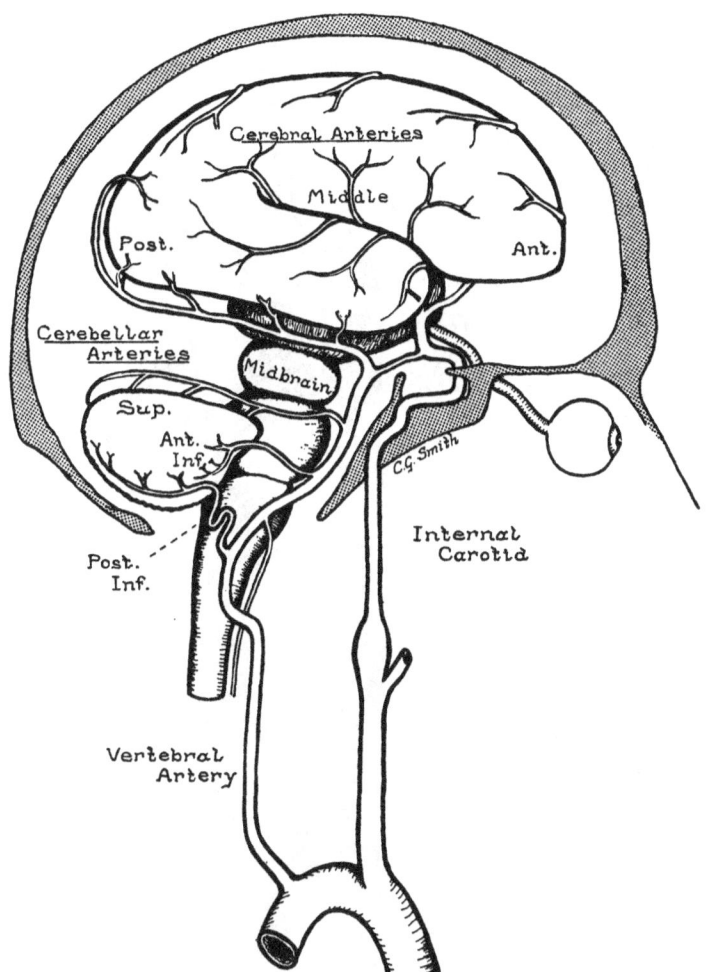

FIGURE 112. The source and general distribution of the large arteries that enter the skull to supply the brain.

fellow of the opposite side at the midline at the lower border of the pons (figure 113). The common vessel so formed is the **basilar artery**. A median septum is occasionally seen in the basilar as evidence of its formation from right and left parallel vessels. The basilar is as long as the pons segment. It ends by dividing into right and left branches that are called the **stems of the posterior cerebral arteries**. These parts of the right and left ventral arterial trunks are so named because

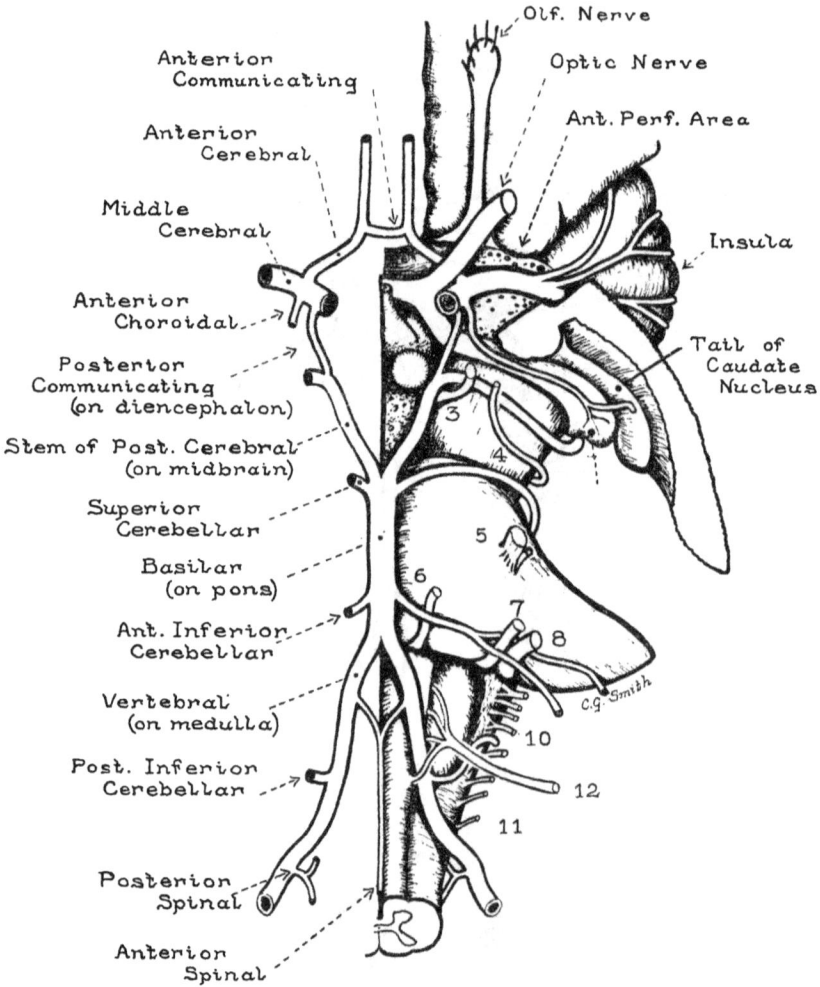

FIGURE 113. The right and the left ventral arterial trunks of the brain, their anastomoses with each other and their large branches.

they carry most of the blood that is distributed by the posterior cerebral arteries. They ascend at the borders of the interpeduncular fossa as far as the diencephalon. There the slender **posterior communicating artery**, the diencephalic portion of the longitudinal arterial trunk, leaves the end of the stem of the posterior cerebral to connect it with the internal carotid. The part of the posterior cerebral artery that extends laterally from this site is an enlargement of a side branch of the ventral trunk. It courses laterally around the side of the brain stem. In atypical cases the posterior cerebral receives most of its blood from the internal carotid instead of the basilar. In such cases the diameter of the posterior communicating artery is correspondingly large and serves as the stem of the posterior cerebral.

The **internal carotid artery**, which delivers blood into the upper end of the arterial trunk, enters the base of the skull on the lateral side of the pharynx and then courses medially in the bony roof of the pharynx, that is, in the temporal bone, to reach the foramen lacerum. Through this foramen it ascends into the posterior end of the cavernous sinus at a point just lateral to the dorsum sellae (figure 112). It courses forward on the floor of the sinus to its anterior wall. Here it makes a right-angled turn and ascends to pierce the dural roof of the sinus and strike the under side of the optic nerve just behind the optic foramen. The optic nerve deflects it, causing it to turn posteriorly and laterally, to the side of the optic chiasma where it is able to continue its ascent to the anterior perforated area (figure 113). There it divides into its terminal branches, the **anterior** and **middle cerebral arteries**. The anterior cerebral courses forward and medially above the optic nerve toward the midline just in front of the lamina terminalis. The middle cerebral artery courses laterally and posteriorly onto the insula.

The parts of the internal carotid that lie respectively on the floor, the anterior wall, and on the roof of the cavernous sinus are considered by the radiologist to be a unit. He calls it the **carotid siphon**.

III. THE BRANCHES OF THE VENTRAL ARTERIAL TRUNK

A. THE CEREBELLAR ARTERIES

The cerebellar arteries are named according to the surfaces of the cerebellum that receive them. There is one for the superior surface and there are two for the inferior surface, an anterior and a posterior.

1. THE SUPERIOR CEREBELLAR ARTERY

This artery arises from the basilar at the upper border of the pons and passes laterally and dorsally hugging the isthmus region of the pons segment (figures 113, 114). On the dorsal surface of the pons

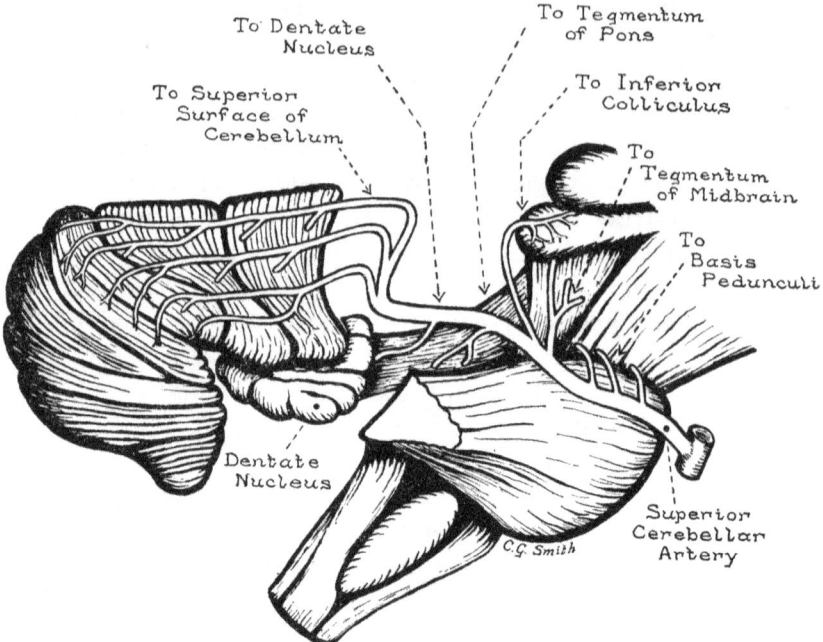

FIGURE 114. The course and the distribution of the superior cerebellar artery.

segment where it is overlapped by the cerebellum, it gives off its cortical branches, two to four of them. These bend around the anterior border of the tentorial surface of the cerebellum and course backward in the sagittal plane. They reach as far as the inferior semilunar lobule where they anastomose with the terminal branches of the inferior cerebellar arteries.

In addition to its terminal branches the artery has important collateral branches. Near its origin it gives off very fine branches that sweep laterally and penetrate and supply the basis pedunculi where this fibre bundle enters the pons. At the dorsolateral border of the basis pedunculi and beyond this point it gives off additional fine branches, some of which penetrate the brachium conjunctivum to supply it and the tegmentum of the pons. Others penetrate the lateral lemniscus and supply the tegmentum of the midbrain. One or two relatively large branches accompany the brachium conjunctivum into the core of the cerebellum to supply the dentate nucleus. Another relatively large branch courses onto the inferior colliculus to supply this part of the tectum of the midbrain.

2. THE ANTERIOR-INFERIOR CEREBELLAR ARTERY

This arises from the basilar close to the inferior border of the pons segment (figure 115). It courses laterally toward the flocculus usually in front of nerve 6 so as to press it against the pons. Before reaching the flocculus it divides into two branches, an upper and a lower. The upper branch follows the brachium pontis and turns onto the inferior surface of the cerebellum lateral to the flocculus. The inferior branch usually passes laterally below and in contact with the seventh and

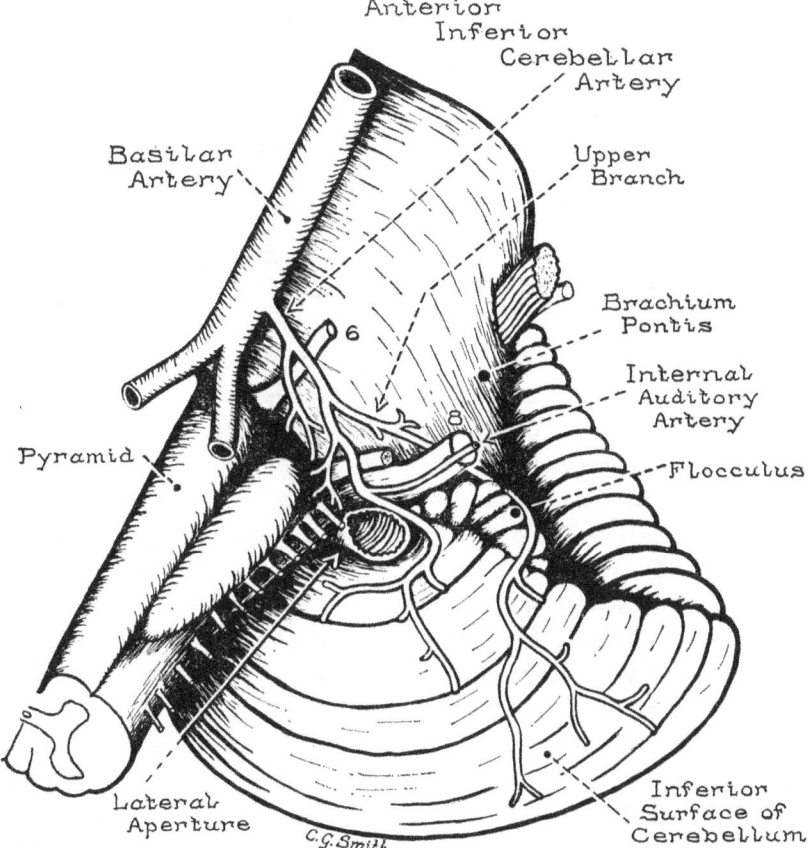

FIGURE 115. The course and distribution of the anterior-inferior cerebellar artery.

eighth nerves to reach the inferior surface of the cerebellum medial to the flocculus. As it crosses the eighth nerve it usually gives off the internal auditory artery that follows that nerve to the sense organs for hearing and position in the internal ear. The internal auditory artery may come directly from the basilar artery.

The non-cortical branches of the anterior-inferior cerebellar artery are (1) the internal auditory, (2) nutrient branches that pierce the caudal part of the pons and the brachium pontis, (3) the nutrient branches that enter the medulla oblongata along the caudal border of the pons lateral to the pyramid. These nutrient vessels to the pons and medulla penetrate as far as the floor of the fourth ventricle.

3. THE POSTERIOR-INFERIOR CEREBELLAR ARTERY

This vessel does not come from the basilar artery as it should but from the vertebral (figures 113, 116). Moreover, it leaves the vertebral near the caudal end of the medulla. One would expect this artery to reach the cerebellum along its stalk and it does, but only after it has executed a preliminary, S-shaped bend in order to supply nutrient branches to the lateral and dorsal aspects of the medulla.

As figure 116 shows, the artery turns around the caudal end of the olive to ascend ventral to the rootlets of nerves 11 and 10. It does this to deliver a leash of fine vessels that enter the medulla between the olive and the restiform body. The posterior-inferior cerebellar artery then turns dorsally, usually about the middle of the olive, passing through the rootlets of nerves 11 or 10 to reach the cuneate tubercle. Here it turns caudally and gives off fine branches that enter the cuneate tubercle. Its descent carries it well below the obex where it makes a hair-pin turn and ascends on the medulla and skirts the side of the inferior aperture to course up onto the roof of the fourth ventricle. In this part of its course it gives off a series of very fine vessels closely spaced like the teeth of a comb that penetrate the medulla along the attachment of the inferior velum. Fine twigs also pass directly to the choroid plexus of the fourth ventricle.

When the artery approaches the stalk of the cerebellum it divides into two terminal cortical branches, a medial one for the inferior surface of the vermis and a lateral one for the medial part of the inferior surface of the cerebellar hemisphere. These sweep backwards

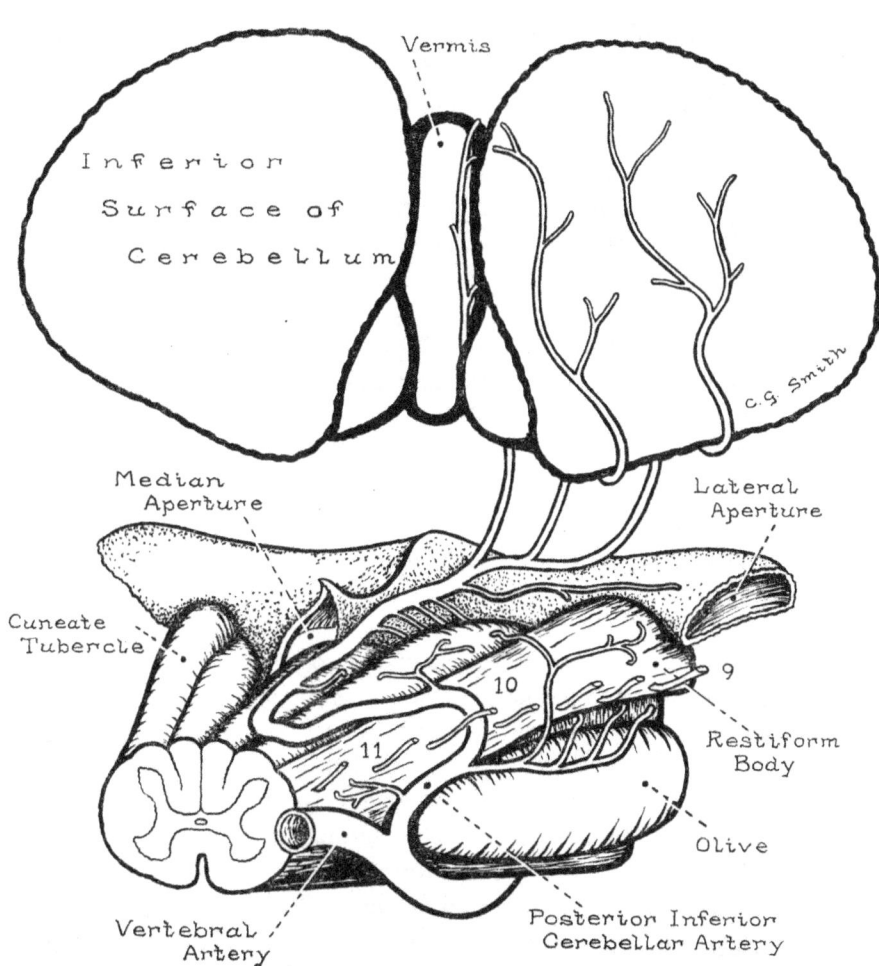

FIGURE 116. The course and distribution of the posterior-inferior cerebellar artery.

clinging to the cerebellum and reach as far as the tuber and the inferior semilunar lobule respectively.

The course of the artery as given above conforms to this description in about 70 per cent of cases. In about 10 per cent the anterior-inferior cerebellar enlarges its distribution to supply the whole inferior surfaces. In some cases the posterior-inferior cerebellar artery passes almost directly backwards from the vertebral omitting its ascent on the side of the medulla. In such cases the lateral side of the medulla plus the upper ends of the restiform body and the cuneate tubercle are supplied by a series of side branches of the vertebral artery.

B. THE CEREBRAL ARTERIES

The cerebral hemisphere, like each cerebellar hemisphere, is supplied by three arteries. The cerebral arteries are named anterior, middle, and posterior.

1. THE ANTERIOR CEREBRAL ARTERY

This leaves the internal carotid as it ascends lateral to the chiasma (figure 117). It courses medially and anteriorly between the optic nerve below and the anterior perforated area above, to follow the medial olfactory stria onto the medial surface of the hemisphere. This brings the right and left anterior cerebral arteries to within 1 or 2 mm. of each other and at this site they are connected by the short, stout anterior communicating artery.

On the medial surface of the hemisphere the anterior cerebral artery courses directly to the genu of the corpus callosum and then clings to the corpus callosum as far back as the splenium. Its cortical branches, some larger than the parent vessel, come off at irregular intervals after it passes onto the medial surface of the hemisphere. These extend obliquely toward the margin of the hemisphere and continue around it onto the orbital surface and onto the lateral surface. They supply the cortex of the medial surface to within 1 cm. of the parieto-occipital fissure and on the orbital and lateral surfaces they supply an adjacent marginal band of cortex about 3cm. wide. This marginal band extends back to within 1 cm. of the occipital lobe.

The stem of the anterior cerebral artery gives off fine nutrient vessels throughout its course. Close to its origin it sends branches to the optic chiasma and to the preoptic region of the hypothalamus. As it approaches the medial border of the hemisphere it sends branches through the anterior perforated area to supply the head of the caudate, the anterior pole of the lentiform nucleus, and the anterior limb and genu of the internal capsule. These are the striate branches of the anterior cerebral. One of these is much larger than the rest and it delivers most of the blood to the above portion of the striate region. It begins close to the anterior communicating artery

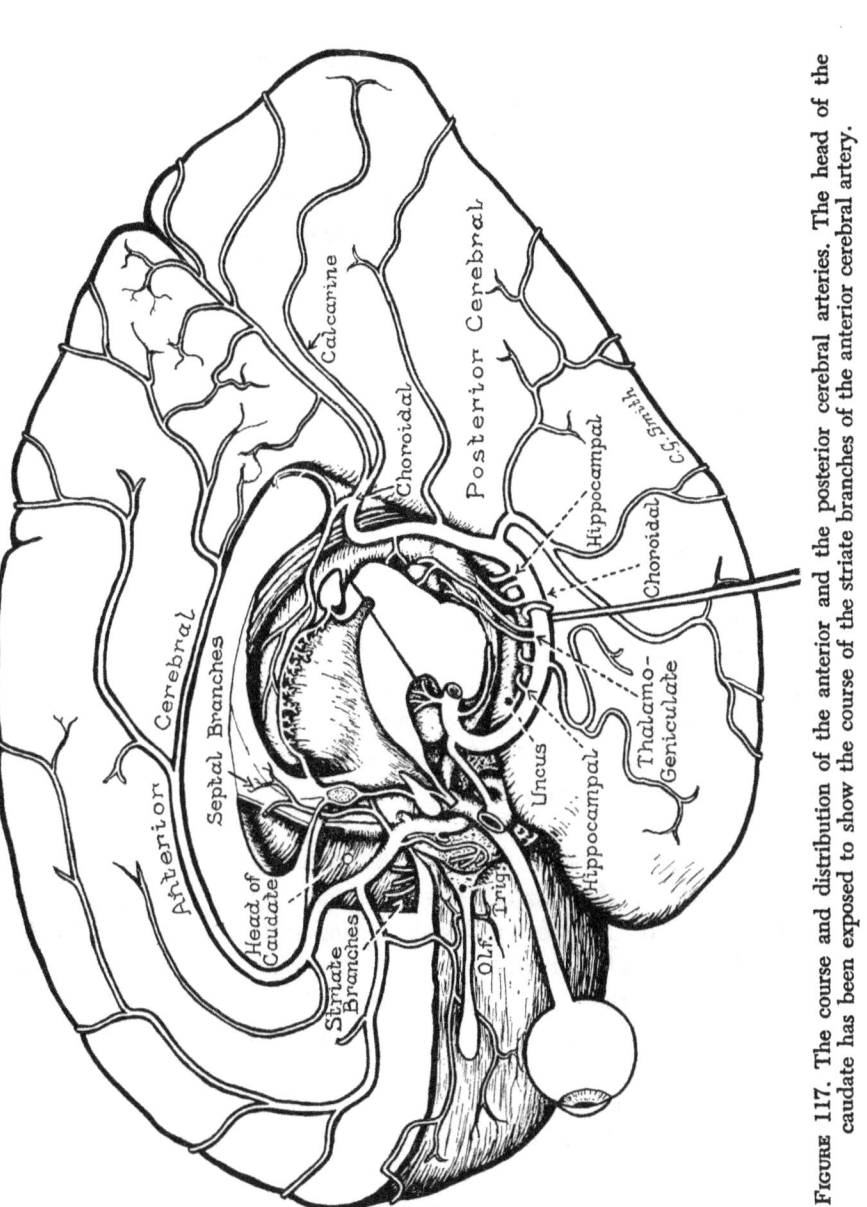

FIGURE 117. The course and distribution of the anterior and the posterior cerebral arteries. The head of the caudate has been exposed to show the course of the striate branches of the anterior cerebral artery.

at the medial border of the hemisphere and courses back to reach the base of the olfactory trigone. This artery was first described by Heubner in 1872 and commonly is referred to by his name. It divides into three or four smaller vessels that enter foramina located behind the lateral part of the base of the olfactory trigone. In figure 117 a part of the septal portion of the medial wall of the hemisphere has been removed to show the striate vessels entering the head of the caudate nucleus.

As the anterior cerebral begins its course on the medial surface it gives off branches that supply the septal area including the septum pellucidum. As it lies on the corpus callosum it gives off a series of fine nutrient vessels that supply all but the splenium (supplied by the posterior cerebral).

From a practical standpoint the distribution of the anterior cerebral has the following significant features. (1) It supplies the cortex of the motor area of the lower limb and the motor pathway to the head and arm in the internal capsule. (2) It supplies the septal region where small lesions may induce a prolonged state of unconsciousness. (3) It supplies the corpus callosum which has in it the significant pathway from the dominant hemisphere to the motor area of the non-dominant hemisphere. The functional deficit that follows the interruption of this pathway is known as apraxia.

2. THE MIDDLE CEREBRAL ARTERY

The middle cerebral artery begins lateral to the chiasma (figure 118). It is one of the two terminal branches of the internal carotid. It courses laterally below the anterior perforated area to the limen insulae. As it approaches the insula it begins to divide into a leash of cortical branches. These fan out on the insula and then pass onto the deep surface of its opercula to reach the borders of the lateral fissure. From the lateral fissure the branches extend to within 3 cm. of the superior margin, to within 1 cm. of the inferior margin of the temporal lobe, and as far back as the occipital lobe. They also extend onto the lateral two-thirds of the orbital surface and onto the temporal pole but not to the uncus which is supplied by the anterior choroidal artery. The vessels are shown in figure 118 as they were seen in one injected specimen. Note that one of these enters the central sulcus to supply the portion of the motor area and the portion of the general sensory area of the upper half of the body.

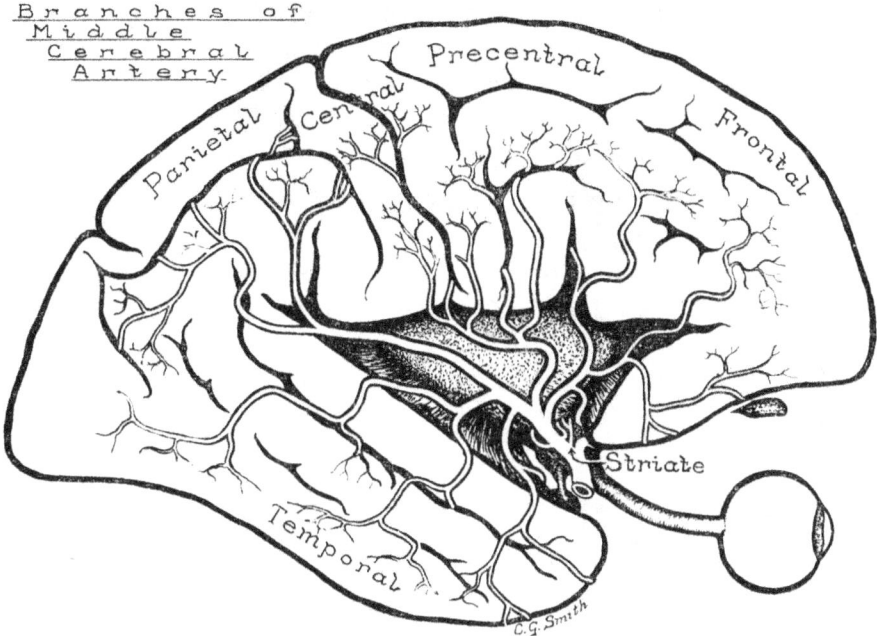

FIGURE 118. The course and distribution of the middle cerebral artery. The opercula are pulled apart to expose the striate branches and the beginning of each of the cortical branches on the insula.

The stem of the middle cerebral artery gives off nutrient vessels that enter the brain through the anterior perforated area. These are the striate branches of the middle cerebral artery. The largest branches are crowded together in the lateral part of the perforated area where they find their way into the lower edge of the external capsule. They fan out within this lamina clinging at first to the lateral aspect of the lentiform nucleus but gradually penetrating it. The largest of these vessels can be followed by dissection (figure 120). They supply (1) all the putamen and all the caudate except in each case the frontal ends which are supplied by the anterior cerebral; (2) the lateral part of the globus pallidus; (3) all that part of the anterior limb, the genu, and the posterior limb, that lies adjacent to the putamen; (4) the optic radiation in the anterior part of the core of the temporal lobe. This supply to the optic radiation was first described by Abbie in 1937. The vessel involved cannot always be seen in the dissection of uninjected specimens but when present it forms the core of the optic radiations.

234   BASIC NEUROANATOMY

If it were absent it would be replaced presumably by branches of the cortical arteries on the surface of the temporal lobe.

The crowding of the nutrient vessels into the lateral angle of the perforated area is due to the buckling of its lateral border as a result of the growth forward of the elongating temporal lobe.

3. THE POSTERIOR CEREBRAL ARTERY

The right and the left posterior cerebral arteries are the terminal branches of the basilar (figures 117, 119). Each one ascends at the border of the interpeduncular fossa medial to the third nerve and then

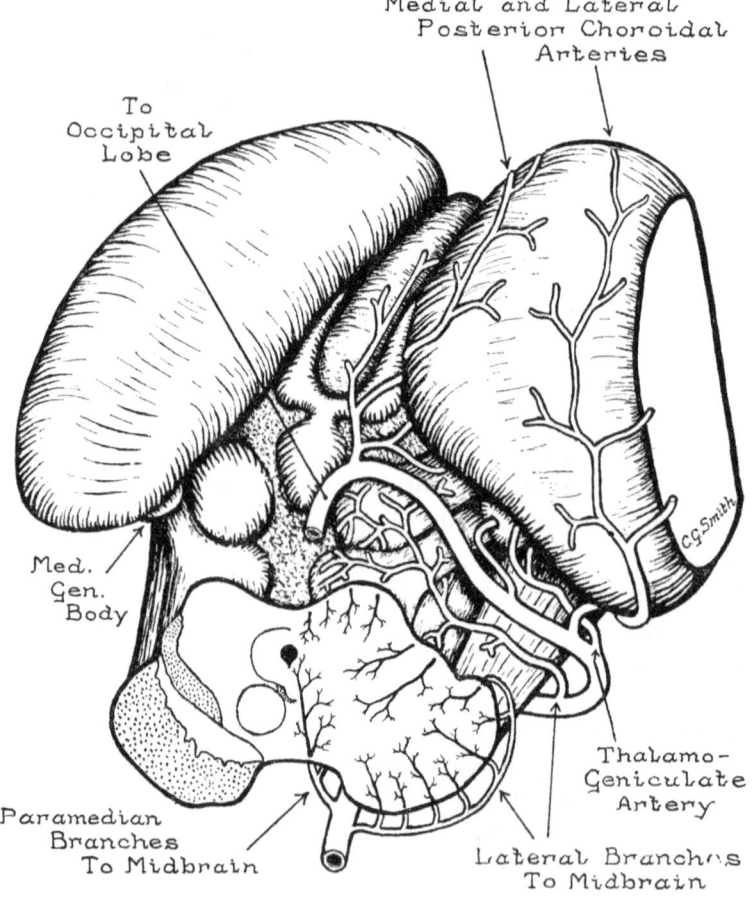

FIGURE 119. The branches of the posterior cerebral artery that supply the midbrain and the diencephalon.

turns laterally around it as around a marker buoy (Figure 113). As it courses laterally (figure 117) it lies in the cleft between the parahippocampal gyrus and the brain stem. This conducts it to a point just below the splenium where it makes a right-angled turn and crosses the end of the parahippocampal gyrus to enter the end of the calcarine sulcus. Its terminal branches lie in the parieto-occipital and the posterior calcarine sulci respectively. The distribution of the calcarine branch coincides precisely with the visual area of the cortex. Branches of the parieto-occipital artery supply the rest of the cuneus, plus a strip of cortex 1 cm. wide along the anterior border of the parieto-occipital fissure and they cross the superior border to help supply the lateral surface of the occipital lobe.

The other cortical branches of the posterior cerebral are collateral branches that leave it as it courses along the medial border of the parahippocampal gyrus. There are two sets of these, one set for the neocortex and one for the archicortex of the hippocampal formation.

The arteries to the neocortex, three or four of them, are large and course laterally across the parahippocampal gyrus fanning out on the tentorial surface of the temporal and the occipital lobes. At the inferolateral border of the hemisphere they turn onto the lateral surface and supply the inferior temporal gyrus and the lower part of the occipital lobe.

The branches that supply the cortex of the hippocampal formation are short and slender. They pass to the medial surface of the parahippocampal gyrus and at the hippocampal sulcus they turn to run along the free surface of the dentate gyrus. In the specimen illustrated in figure 117, two hippocampal arteries entered the uncal portion of the hippocampal fissure and two entered the fissure near its midpoint. As these arteries course along the hippocampal sulcus they give off nutrient vessels to the dentate gyrus and other longer, penetrating vessels that reach the hippocampus proper in the hippocampal eminence of the lateral ventricle. It is the longer vessels to the hippocampus that are most likely to be occluded by pressure caused by a herniation (protrusion) of the hippocampal gyrus into the interval between the midbrain and the free edge of the tentorium. Hence the hippocampus is damaged more than the dentate gyrus in such a condition. The pathologist calls the vulnerable part of the cross-section of the hippocampal formation Sommer's sector.

The non-cortical branches of the posterior cerebral supply the choroid plexus of the lateral ventricle, the choroid plexus of the third ventricle, diencephalon, and the midbrain.

A lateral posterior choroidal artery leaves the posterior cerebral opposite the lateral geniculate body (figures 117 and 119). This courses forward into the lateral part of the transverse fissure on the lateral part of the thalamus as far as the anterior tubercle. It sends some branches to the choroid plexus of the lateral ventricle. Others cling to the lateral part of the thalamus to supply it through its dorsal surface. A medial posterior choroidal artery leaves the posterior cerebral near the midline just below the splenium. It courses forward into the transverse fissure also (but close to the midline) and its branches supply the choroid plexus of the third ventricle, the epithalamus, and the medial part of the thalamus through its dorsal surface.

The posterior cerebral, as we shall see, supplies most of the diencephalon. It supplies all the dorsal portion of the diencephalon through branches of its choroidal arteries. It supplies most of the inferior portion of the diencephalon by direct branches that enter its inferior surface. Of these, three or four, called the thalamogeniculate arteries, leave the posterior cerebral lateral to the basis pedunculi and enter the diencephalon through the metathalamus (figure 119). An additional cluster of perforating vessels, called the thalamoperforating vessels, enters the diencephalon just behind the mammillary bodies, that is, through the subthalamus (figure 117).

Since the stem of the posterior cerebral is, developmentally, the midbrain portion of the ventral arterial trunk, it follows that it should supply the midbrain and it does. The branches to the midbrain are side branches that extend medially and laterally. The medial ones enter the midbrain through the posterior perforated area. These paramedian vessels are relatively large and long. They penetrate as far as the aqueduct supplying in turn the medial part of the basis pedunculi, the medial parts of the brachium conjunctivum and red nucleus, and the motor nuclei of the third and the fourth nerves (figure 119).

The lateral branches of the posterior cerebral to the midbrain extend across the basis pedunculi toward the dorsal surface of the midbrain. Usually only one vessel reaches the tectum. It branches to supply the superior and inferior colliculi. All these vessels give off branches that penetrate the basis pedunculi or the lemnisci or the brachium of the inferior colliculus to supply the portion of the tegmentum that is not supplied by the paramedian vessels.

C. THE ANTERIOR CHOROIDAL ARTERY

The name of this artery does not do justice to its distribution (figures 113, 120, 121). It supplies part of the choroid plexus of the lateral ventricle but its important distribution is to the optic, pyramidal, and pallidothalamic pathways.

The artery accompanies the optic tract. It comes from the internal carotid at the lateral side of the optic chiasma and clings to the optic tract to reach the lateral side of the lateral geniculate body, where it enters the choroid plexus of the inferior horn. Its branches to the optic pathway are a series of collateral branches that are distributed to the optic tract, the lateral part of the lateral geniculate body and, by penetrating this, the optic radiation in the sublenticular part of the internal capsule. Its branches to the pyramidal pathway are another set of collateral branches that are distributed to the posterior limb of the internal capsule immediately above the basis pedunculi. (This part of the posterior limb is applied to the globus pallidus.) These nutrient vessels, like the teeth of a comb, enter the lower border of the posterior limb chiefly along the lateral border of the optic tract but some pierce it.

Its branches to the pallidosubthalamic pathway accompany the branches to the posterior limb. Of these the lateral ones supply the globus pallidus; the medial ones penetrate the junction of basis pedunculi and posterior limb to reach the subthalamic nucleus.

Developmentally the anterior choroidal artery is a fourth cerebral artery for the supply of the posterior part of the olfactory pathway to the cerebral cortex. Hence it has branches that arborize on the surface of the uncus and enter the hippocampal sulcus to help the posterior cerebral artery supply the hippocampal formation.

D. THE PONTINE BRANCHES OF THE BASILAR ARTERY

The pontine branches of the basilar are a series of vessels that supply the pons through its ventral and lateral surfaces. About four to six of these enter the pons close to the midline on each side. These paramedian vessels like those of the midbrain are long and supply a paramedian region that contains (a) a portion of the motor pathway and (b) a nucleus of the somatic motor column, namely, the abducens nucleus. Another four to six pontine vessels course laterally onto the brachium pontis. These give off nutrient vessels that penetrate and supply the lateral portion of the basilar part of the pons and the tegmentum as deep as the fourth ventricle. Their distribution will be discussed later in dealing with the blood supply of the pons segment.

E. THE ANTERIOR SPINAL ARTERY

This artery leaves the medial side of the vertebral near its upper end and courses medially to unite with its fellow of the opposite side and descend in the midline onto the spinal cord (figure 113).

F. THE DORSAL SPINAL ARTERY

This artery is smaller than the anterior spinal and comes off the lateral side of the vertebral at the caudal end of the medulla (figure 113). It courses dorsally onto the fasciculus cuneatus where it divides into a short, ascending and a long, descending branch. The descending branch has a zig-zag course on the dorsal surface of the cord. Its branches anastomose freely with its fellow of the opposite side and also with branches of the anterior spinal (figure 123).

IV. THE BLOOD SUPPLY OF THE INTERNAL CAPSULE

The blood supply of the internal capsule is illustrated in figure 120. To understand the distribution of vessels to the internal capsule it is necessary to keep in mind that the lateral surface of the anterior limb and of the posterior limb faces not only laterally but also inferiorly. Hence in a coronal section of the hemisphere the inferior border of the internal capsule will be nearer the midline than the superior border. Since nutrient vessels ascend almost vertically through perforations on the inferior surface of the brain it follows that the superior-lateral part and the inferior-medial part of the capsule will be supplied by different nutrient arteries. Let us consider the blood supply of each of the parts of the internal capsule in turn.

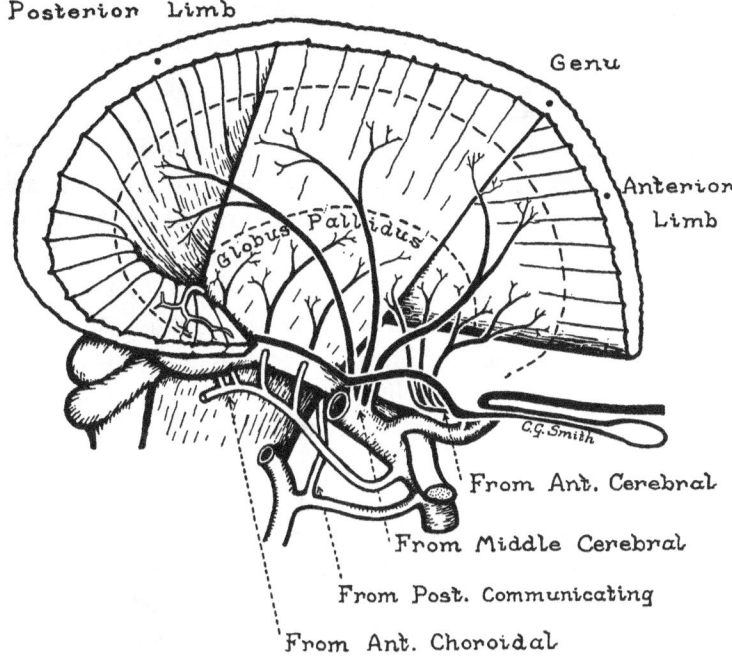

FIGURE 120. The lateral aspect of the internal capsule to show its blood supply. The two concentric broken lines indicate the position of the borders of the lentiform nucleus and the globus pallidus respectively.

A. THE ANTERIOR LIMB

This part of the capsule is supplied by the striate branches of the anterior cerebral and the middle cerebral. The anterior cerebral supplies its inferior-medial part and the middle cerebral supplies its superior-lateral part. These nutrient vessels reach their destination through the anterior perforated area.

B. THE GENU

The genu is also supplied by striate branches of the middle and anterior cerebral arteries and in the same way as the anterior limb. However it has an additional blood supply from branches to the diencephalon. These come from the posterior communicating artery or occasionally directly from the internal carotid. They enter the base of the brain just lateral to the mammillary body and course along the medial side of the genu supplying it and the adjacent part of the diencephalon as far as the anterior nucleus of the thalamus.

C. THE POSTERIOR LIMB

The posterior limb receives striate branches from the middle cerebral and from the anterior choroidal arteries. The branches from the middle cerebral enter the external capsule through perforations in the lateral part of the perforated area. They fan out in this lamina and branch and turn into the putamen of the lenticular nucleus to reach the portion of the posterior limb that extends above the globus pallidus. The branches from the anterior choroidal enter the brain through the optic tract or along its borders as it lies on the basis pedunculi. Some of these pass directly into the posterior limb; others reach it after passing through the globus pallidus.

D. THE SUBLENTICULAR "LIMB"

The sublenticular part of the internal capsule is supplied by the anterior choroidal artery. Its branches penetrate the lateral border of the optic tract and the lateral part of the lateral geniculate body and pass directly into the optic radiations.

V. THE BLOOD SUPPLY OF THE DIENCEPHALON

Nutrient vessels of the diencephalon are restricted to its pia-covered surfaces, that is, to its superior and inferior surfaces (figure 121). The nutrient vessels of the superior surface come from the posterior choroidal branches of the posterior cerebral. They are uniformly distributed and enter the diencephalon at right angles to the surface. They reach a third of the distance to the inferior surface.

The nutrient vessels of the inferior surface come directly from large basal arteries. They are not uniformly distributed because those that should enter where the midbrain is attached are displaced, some forward and some backward. These displaced vessels have an arched course to reach their destination in the part of the diencephalon directly above the attachment of the midbrain.

The chief source of nutrient vessels to the inferior part of the diencephalon is an arterial trunk consisting of the posterior communicating and the posterior cerebral arteries. This trunk extends from the optic

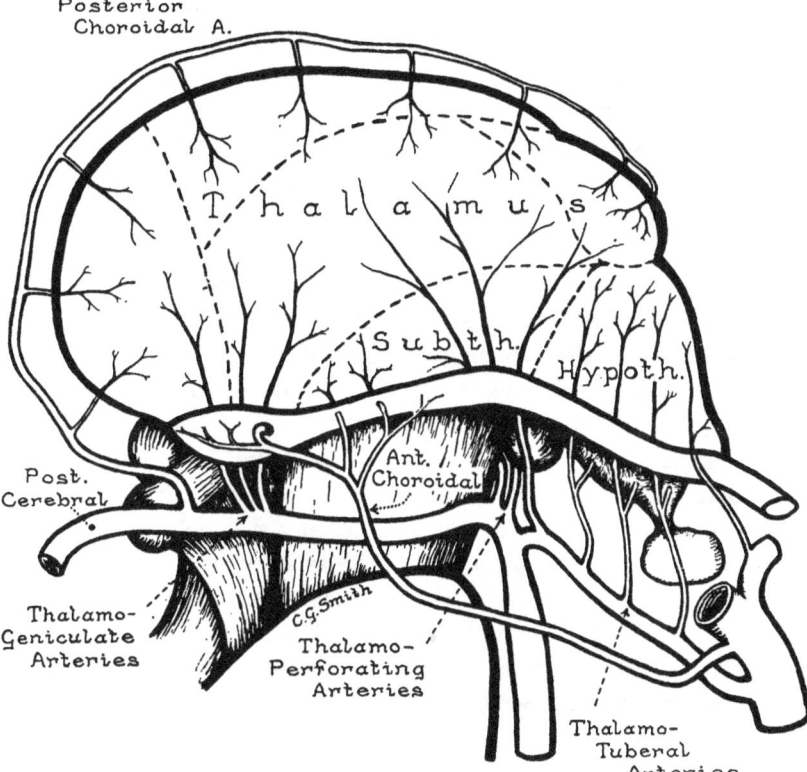

FIGURE 121. The lateral aspect of the diencephalon and the midbrain to show the blood supply of the subthalamus, the hypothalamus, the lateral and medial geniculate bodies, and the thalamus.

chiasma around the side of the basis pedunculi to the pulvinar. Its nutrient branches are in three groups: two of these—the thalamo-tuberal and the thalamoperforating—enter the diencephalon in front of the attachment of the midbrain, the third group—the thalamogeniculate vessels—enter the diencephalon behind the attachment of the midbrain.

The thalamotuberal vessels leave the posterior communicating artery as it lies alongside the hypothalamus. They enter the tuberal region of the hypothalamus and penetrate into the overlying nuclei of the thalamus, hence their name. They supply all but the anterior end of the hypothalamus which receives fine vessels from the internal carotid and the anterior cerebral arteries.

The thalamoperforating vessels are so named because they reach the thalamus through the posterior perforated area. They are long, stout vessels that leave the posterior communicating artery and the stem of the posterior cerebral and enter the subthalamus lateral to the mammillary body and also immediately behind it. They diverge in the subthalamus and reach into the nuclei of the overlying thalamus. These vessels supply all the subthalamus except its most lateral part, which receives nutrient vessels from the anterior choroidal. The latter pierce the fibres of the basis pedunculi just deep to the optic tract.

The thalamogeniculate vessels are so named because they enter the thalamus through the geniculate bodies. They diverge as they enter the diencephalon and help to supply the thalamic nuclei above the attachment of the midbrain. In passing through the geniculate bodies these vessels supply them, except for the lateral half of the lateral geniculate body which is supplied by the anterior choroidal artery.

The thalamogeniculate vessels are the ones that help to supply the nucleus ventralis posterior, which is the relay station for the general sensory pathway. Occlusion of these vessels will therefore be followed by some loss of sensation on the opposite side of the body. The condition is characterized usually by spontaneous, excruciating pain referred to the side of the body where the interrupted pathway begins.

VI. THE BLOOD SUPPLY OF THE PONS SEGMENT

The pons segment is described as having a ventral paramedian set of nutrient vessels and a lateral set. The paramedian set enter the pons close to their origin from the basilar. Of these, the ones leaving the middle part of the basilar artery have to penetrate the thickest part of the pons and manage to reach only into the marginal part of the tegmentum. Toward the upper and the lower borders of the pons the thickness of the basilar portion gets progressively less and the corresponding upper and lower perforating nutrient vessels are able to penetrate deeper into the tegmentum. These upper and lower paramedian vessels not only penetrate progressively deeper into the tegmentum as the pons gets less massive, they also alter their course to arch toward each other in the sagittal plane and supply the intervening portion of the tegmentum.

The lateral part of the tegmentum is only partially covered by the pons and by its lateral extension, the brachium pontis. In the interval between the brachium pontis below, and the midbrain above, the nutrient vessels can enter the tegmentum directly from the superior cerebellar artery and course medially to penetrate as far as the paramedian region. Where the pons covers the lateral part of the tegmentum it is supplied like the paramedian portion of the tegmentum by nutrient vessels that arch into it from above and below, that is, from the vicinity of the midbrain and from the vicinity of the medulla oblongata. Those reaching it from above are branches of the superior cerebellar artery and branches of the lateral pontine vessels. Those reaching it from below are branches of the anterior-inferior cerebellar artery and branches of the lateral pontine vessels.

VII. THE BLOOD SUPPLY OF THE MEDULLA OBLONGATA

The nutrient vessels of the medulla oblongata fall into two groups, paramedian and lateral (figure 122). The paramedian vessels enter the medulla through its ventral surface, that is, medial to the olive. They are long vessels that penetrate the pyramid, then the medial

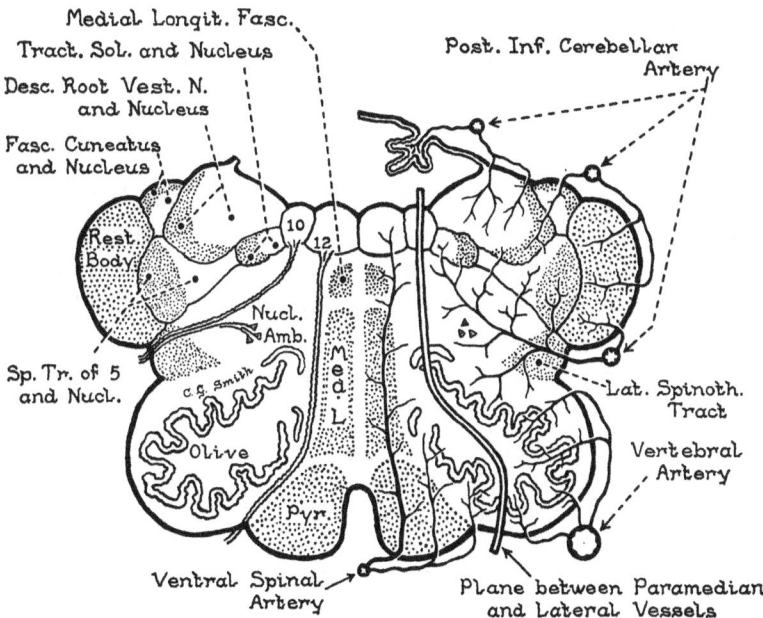

FIGURE 122. A cross-section of the medulla oblongata to show its blood supply.

lemniscus, and end in the motor nuclei in the floor of the fourth ventricle. Near the pons these vessels come from the vertebral artery but caudal to the origin of the ventral spinal artery they come from the latter vessel.

The nutrient vessels of the lateral and dorsal portions of the medulla are all short vessels except for those that enter the medulla in the depression lateral to the olive. Some of these long vessels penetrate to the medial lemniscus, others extend dorsally to supply the sensory nuclei in the floor of the fourth ventricle. The source of these vessels shifts from one artery to another throughout the length of the medulla. Near the upper pole of the olive they come from the anterior-inferior cerebellar artery; below this level they come from the vertebral and the posterior inferior cerebellar arteries. Where the posterior and the anterior-inferior cerebellar arteries fail to contribute their quota of nutrient vessels, the vertebral makes up the deficit.

The short nutrient vessels of the dorsal aspect of the medulla oblongata supply the nucleus gracilis, the nucleus cuneatus, the caudal part of the vestibular nuclei, and the restiform body (figure 122). These vessels come from the posterior spinal and the posterior-inferior cerebellar arteries. Near the pons where the inferior velum makes up the whole of the dorsal surface of the medulla, the restiform body is supplied by nutrient vessels that enter its ventral lateral border.

VIII. THE BLOOD SUPPLY OF THE SPINAL CORD

The spinal cord receives its blood through the branches of three longitudinal arterial trunks (figure 123). Each of these trunks extends from the brain to the filum terminale. One is a median vessel located in front of the ventral median fissure. The other two, one on the right side of the cord, the other on the left, follow a zig-zag course in the vicinity of the dorsolateral sulcus. The two dorsal trunks communicate with each other through numerous anastomatic channels. Each of these trunks begins on the medulla as a branch of the vertebral artery. As they descend they receive feeder vessels from some of the arteries that reach the cord along the nerve roots.

The feeder vessels for the ventral or anterior arterial trunk reach it along four to six of the ventral roots. Of these some come in on the right side, others on the left. Similarly each of the dorsolateral trunks

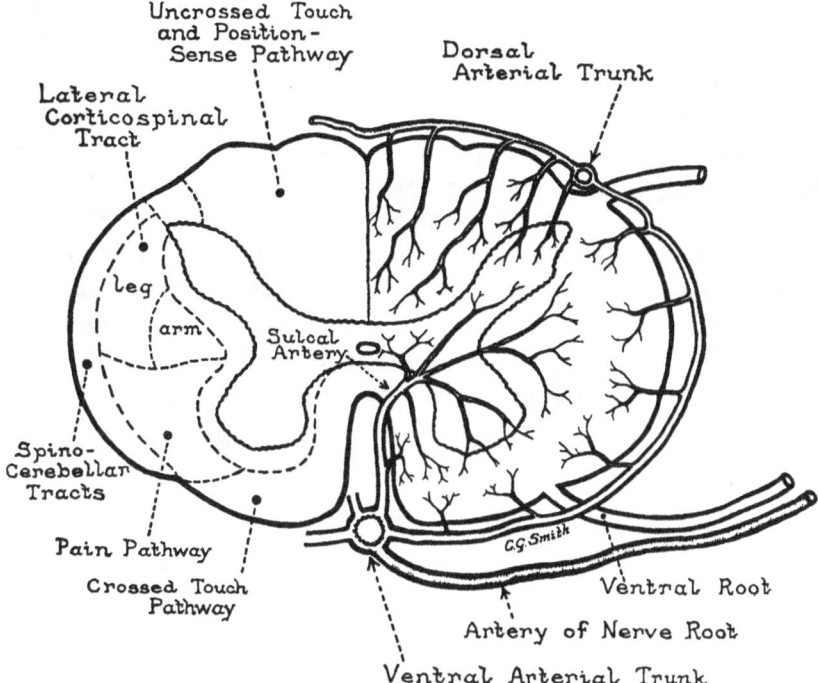

FIGURE 123. A cross-section of the spinal cord to show the course and distribution of its nutrient vessels.

receives only two or three feeder vessels. The feeder vessels are not uniformly spaced along the length of the cord, nor is the pattern consistent from specimen to specimen. Nevertheless anatomical studies and studies of pathological material indicate that the arterial trunks of the cord receive most of their blood through vessels reaching the cervical and the lumbar segments. The available data suggest that blood flows caudally, in the upper part of the trunks, to about the fourth thoracic segment and ascends in the lower part of the cord to the same segment. Hence the fourth thoracic segment is prone to suffer most when the amount of blood reaching the cord is reduced through injury or disease. Such a reduction in the flow of blood to the cord could follow an injury to a nerve root if it happened to be conveying a large feeder vessel.

The nutrient vessels of the spinal cord. The largest nutrient arteries of the cord are the sulcal vessels (figure 123). These arise from the ventral spinal arterial trunk and pass back to the floor of the ventral median sulcus. There each sulcal vessel turns to one side or the other and enters the cord to branch and supply all the grey matter except the dorsal one-third of the dorsal horn and all the fibres of the ventral and the lateral funiculus except for a thin, superficial layer. This artery therefore supplies both the lateral and the ventral corticospinal tracts. The rest of the fibres of the ventral and lateral funiculi are supplied by short, perforating branches from the pial network of vessels formed by branches of the ventral arterial trunk and the dorsolateral one.

The dorsal funiculus and the dorsal portion of the dorsal horn are supplied by nutrient vessels that penetrate the dorsal surface of the cord from the pial network formed by the anastomosing branches of the right and left dorsolateral arterial trunks.

Index

ABDUCENS NERVE. *See* Nerve
Accessory nerve. *See* Nerve
Acoustic nerve. *See* Nerve
Alveus, 197
Agranular cortex, 170
Ala cinerea. *See* Trigone, vagal
Ansa lenticularis, 81
Apertures of fourth ventricle, 32, 38
Aphasia, 184
Apraxia, 232
Aqueduct, cerebral, 11, 54
Archicerebellum, 149
Archicortex
  structure, 167
  location, 180, 187
Area
  of cerebral cortex
    association
      definition, 170
      classification, 183, 184
      lesions and aphasia, 184
      of archicortex, 187
    entorhinal, 187
    motor
      pyramidal, 180, 181
      extrapyramidal, 180, 181
    olfactory, 180
    parolfactory, 179
    prefrontal, 184
    premotor, 181
    sensory
      auditory, 180
      general, 179
      visual, 180
    speech, 184
    thalamic connections, 184
    visual, 180
  perforated, anterior, 157, 199, 201
  perforated, posterior, 236, 242
  septal
    characteristics, 158
    connections, 188
Artery
  basilar, 223
  cerebellar
    anterior inferior, 227
    posterior inferior, 228
    superior, 225
  cerebral
    anterior, 230
    middle, 232, 233
    posterior, 234
  choroidal
    anterior, 237
    posterior, 236
  communicating
    anterior, 230
    posterior, 224, 241
  of Heubner, 232
  internal auditory, 228
  internal carotid, 225
  pontine, 237
  spinal, 238
  striate of
    anterior cerebral, 230
    middle cerebral, 233
  thalamo-geniculate, 236, 242
  thalamo-perforating, 236, 242
  thalamo-tuberal, 241
  vertebral, 222
Auditory apparatus (pathways)
  nerve, 47, 102, 121
  cochlear nucleus, 47, 114, 121
  superior olive, 47, 48, 121
  lateral lemniscus, 46, 47, 48, 51, 121
  inferior colliculus, 51
  brachium of inferior colliculus, 51
  medial geniculate body, 67
  auditory radiations, 92
  auditory cortex, 180
Axon, 1

BAND, DIAGONAL, 203
Basis pedunculi (crus cerebri), 51, 53, 71, 72
Blindness, word, 184
Blood, supply of
  abducens nucleus, 238
  dentate gyrus, 235
  diencephalon, 240

hippocampus, 235, 237
hypophysis, 89
internal capsule, 238
medulla oblongata, 243
midbrain, 236
pons segment, 242
optic radiations, 233, 237
spinal cord, 244
Body
  geniculate
    lateral, 66, 90
    medial, 67, 90
  mammillary, 68, 85
  restiform. See Restiform body
Brachium
  conjunctivum, 44, 58, 79, 95
  colliculus inferior, 51, 56, 67
  colliculus superior, 51, 52, 92
  pontis, 42, 44, 139
Brain stem, 9
Bulb, olfactory. See Olfactory apparatus

CALCAR AVIS, 206
Capsule
  external, 196, 202
  internal
    definition, 69
    subdivisions, 71
    relationships, 71, 199
    pathways
      anterior limb, 72, 98, 99
      genu, 72, 97
      posterior limb, 72, 97
      sublenticular limb, 71, 72, 90, 97
    blood supply, 238
Cardiac plexus, 213
Carotid siphon, 225
Cells
  granule: of cerebellar cortex, 147;
    of cerebral cortex, 167
  Purkinje, 146, 147, 148
  pyramidal, of cerebral cortex, 167
Cerebellum, 127–153
  archicerebellum, 149
  arteries, 225
  cells
    granule, 147
    Purkinje, 146, 147, 148
  core, 128, 138–143
  corpus cerebelli, 150
  cortex
    location, 128
    afferents, 143–145
    structure, 146
    efferents, 148

development, 41–43
fibres of
  mossy, 147
  climbing, 147
fissures
  horizontal, 136
  posterolateral, 137
  primary, 133
flocculonodular lobe, 137
  afferents, 143, 150
  efferents, 148
flocculus, 137, 138, 143
folia, 130
folium of vermis, 135, 145, 152
function, 151–153
hemisphere, 128
inferior medullary velum, 138
lingula, 132
lobes, 149–151
lobules, 129, 132–140
monticulus
  culmen, 133
  declive, 134
neocerebellum, 149
nodule
  location, 137
  afferents, 143
nuclei
  location, 128
  dissection, 138–143
nuclei
  afferents, 147
  efferents, 148
paleocerebellum, 149
paraflocculus, 136
pedicles, 128
peduncles
  superior. See Brachium
    conjunctivum
  middle. See Brachium pontis
  inferior. See Restiform body
pyramis, 136
surfaces, 130, 131
tonsil, 136
  herniation of, 132
tuber, 135, 152
uvula, 136
vallecula, 132
vermis, 128
Cerebral. See also Cerebrum
  aqueduct, 11, 54
  cortex, 166–170, 198
    areas. See Area
    agranular, 170
    archicortex. See Archicortex

granular, 170
layers of, 167–169
neocortex, 167–169
self-exciting circuits, 169, 184
specific afferents, 169
thalamic connections, 184
  hemisphere
association fibres of, 192, 193
attachment to diencephalon, 156
choroid membrane, 158
commissural fibres, 190, 194
development, 154–163
dominance, 184
frontal section, 190
gyri. *See* Gyrus
hilum, 164
lobes, 172
poles, 163
projection fibres of, 68, 190, 197
septal area, 158
sulci. *See* Sulci
surfaces, 163–165
  peduncle, 53
Cerebrospinal fluid, 11
Cerebrum, 53
Chiasma, optic, 66, 74
Chorda tympani, 218
Choroid plexus, 11, 12
  of fourth ventricle, 32
  of third ventricle, 75
  of lateral ventricle, 159, 205
Cingulum, 192
Claustrum, 202
Clava, 31
Cochlear nerve. *See* Nerve, cochlear
Coeliac plexus, 215
Colliculus
  facial, 108
  inferior, 51
  superior, 51, 56
Commissure
  anterior cerebral, 74, 162, 196
  definition, 7
  grey
    of cerebral hemisphere, 159
    of spinal cord, 16
  habenular, 66, 75
  hippocampal, 161, 196
  posterior, 75
  white, of spinal cord, 19, 23, 24
Conjugate movements of eyes
  automatic, 52
  vestibular, 119, 120
Corona radiata, 198
Corpus

callosum, 161, 194–196
cerebelli, 150
striatum, 156, 198–203
Cortex. *See* Cerebellum, Cerebral
Crus cerebri. *See* Basis pedunculi
Culmen monticuli, 130, 133
Cuneus, 179

DEAFNESS, WORD, 184
Declive monticuli, 134
Decussation
  brachia conjunctiva, 58
  definition, 7
  motor (pyramids), 35
  sensory (medial lemnisci), 36, 37
  subthalamic, 81
  tegmental, 56, 57
Dendrite, 1
Dentate
  gyrus, 187, 235
  nucleus. *See* Nucleus
Diagonal band, 203
Diencephalon
  blood supply, 240
  definition, 59
  development, 8
  shape, 59
  subdivisions, 60–64
  surfaces
    inferior, 66
    lateral, 68
    superior, 64
  third ventricle, 72
Dominance of hemisphere, 184

EDINGER-WESTPHAL, NUCLEUS OF, 110
Effectors, 1
Eminence
  collateral, 206
  hippocampal, 206
Emotion, neural mechanism, 189
Entorhinal area, 187
Epithalamus, 64, 99
Equilibrium, role of cerebellum, 151
External capsule, 196, 202
Extrapyramidal. *See* Area, motor;
  Pathways
Eye
  autonomic innervation of, 64, 75, 87, 218
  movements
    conjugate
      frontal lobe, 181
      occipital lobe, 182

superior colliculus, 52
vestibular nuclei, 119

FACIAL NERVE. *See* Nerve
Fasciculus
  cuneatus
    in cord, 18, 19
    in medulla, 31
  definition, 7
  dorsolateral, 19
  fronto-occipital, 193
  gracilis
    in cord, 19
    in medulla, 31
  hypothalamic, 81
  inferior longitudinal, 193
  lenticularis, 81
  medial longitudinal fasciculus, 120
  proprius, 21
  retroflex, 99
  superior longitudinal, 193
  thalamic, 81, 95
  uncinate, 193
Fibres
  arcuate
    dorsal external, 33
    internal, 37
  association of cerebral hemisphere, 190, 192
  climbing, of cerebellum, 147
  commissural, of cerebral hemisphere, 190, 194
  mossy, of cerebellum, 147
  olivocerebellar, 38, 144
  periventricular of diencephalon, 87
  pontocerebellar, 144
    development of, 42, 43
  projection, of cerebral hemisphere, 68, 190, 197
  vestibulocerebellar, 144
  visceral afferent. *See* Visceral
Field H, 81
Fimbria, 197
Fissure
  horizontal of cerebellum, 136
  lateral of cerebral hemisphere, 157
  posterolateral of cerebellum, 137
  primary, 133
  transverse of cerebrum, 159
Flocculonodular lobe. *See* Cerebellum
Flocculus. *See* Cerebellum
Folia, 130
Folium of vermis, 135, 145, 152
Foramen
  interventricular (of Monro), 11, 76

  of Luschka. *See* Aperture, lateral
  of Magendie. *See* Aperture, median
Forceps
  major, 196
  minor, 196
Forebrain, 8
Formation
  hippocampal, 187
  reticular, 123–126
Fornix
  origin, 187
  course, 196, 197
  termination, 85
Fossa, interpeduncular, 51
Fronto-occipital fasciculus, 193
Funiculus
  dorsal, 18, 19
  lateral, 18, 19
  ventral, 18, 19

GANGLION
  autonomic, 4–7, 208
  definition, 7
  ciliary, 110, 218
  geniculate, 102
  impar, 215
  otic, 219
  semilunar (trigeminal), 101
  sensory, 4
  sphenopalatine, 218
  stellate, 215
  submandibular, 219
  sympathetic
    of sympathetic trunk, 215
    of prevertebral plexus, 216
Geniculate body
  lateral, 66, 90
  medial, 67, 90
Geniculocalcarine tract. *See* Optic, radiation
Gennari, line of, 169, 180
Genu
  of corpus callosum, 194
  of internal capsule, 71, 72
    blood supply, 238
Globus pallidus, 201, 202
  blood supply, 233, 237
Glomus of lateral ventricle, 205
Glossopharyngeal nerve. *See* Nerve
Grey
  matter
    central of midbrain, 54
    definition, 7
    of spinal cord, 16
  ramus communicans, 212

Gyrus
    angular, 178, 184
    cingulate (cinguli), 178, 186
    definition, 167
    dentate, 187
        blood supply, 235
    frontal, 178, 179
    fusiform, 179
    lateral olfactory, 166
    lingual, 179
    parahippocampal, 179
    postcentral, 178, 179
    precentral, 178, 181
    subcallosal, 178
    supramarginal, 178
    temporal, 178, 179
        anterior transverse, 180

H FIELD OF SUBTHALAMUS, 81
Habenular
    commissure, 66, 75
    trigone, 66, 99
Haubenfeld, 81
Hearing. See Auditory apparatus
Hemispheres. See Cerebellum, Cerebral
Hilum
    of cerebral hemisphere, 164
    of lateral geniculate body, 66
Hindbrain, 8
Hippocampal commissure, 161, 196
    eminence, 205
    formation, 187
Hippocampus, 187
    blood supply, 235, 237
Hormones, and hypothalamus, 89
Horn, spinal cord, 16
Hypogastric plexus
    inferior, 215
    superior, 215
Hypoglossal nerve. See Nerve
Hypophysis, 89
Hypothalamic sulcus, 76
Hypothalamus, 62, 82–90
    afferents, 90
    dissectable bundles, 85
    efferent pathways, 87–90
    function, 86
    nuclei and regions, 82–85
    relationships, 82

INDUSIUM GRISEUM, 187
Infundibulum, 74
    blood supply, 89
Innominate substance, 201
Insula

development, 157
function, 180
Intermediate mass of thalamus, 76
Intermediolateral nucleus of spinal cord, 18
Intermesenteric plexus, 215
Internal capsule. See Capsule
Interpeduncular fossa, 51
Intersegmental pathways, 21

LAMINA
    medullary, internal of thalamus, 92
    terminalis, 74
Lemniscus
    lateral
        in pons segment, 46, 47, 48
        in midbrain, 51, 56
    medial
        in medulla, 37, 39
        in pons, 47, 115
        in midbrain, 50, 115
        in subthalamus, 79
        in thalamus, 95
Lentiform nucleus. See Nucleus
Leukotomy (lobotomy), 98
Limbic lobe, 188
Limen insulae, 166
Line of Gennari, 169, 180
Lingula of cerebellum, 132
Lissauer's tract, 19
Lobes
    cerebellar 149–151
    cerebral, 172
    flocculonodular. See Cerebellum
    limbic, 188
Lobules
    ansiform, 136
    biventral, 136
    central of cerebellum, 132
    cerebellar, 132–140
        comparative anatomy of, 138
    gracilis, 136
    paracentral, 179
    paramedian. See Lobule, gracilis
    parietal, inferior and superior, 178
    quadrangular of cerebellum, 133
    semilunar of cerebellum, 135, 136
    simple, 134
Luschka, foramen of. See Aperture lateral
Luys, nucleus of. See Nucleus, subthalamic

MAGENDIE, FORAMEN OF. See Aperture, median

INDEX

Mammillary body, 68, 85
Mammillo-tegmental tract. *See* Tract
Mammillo-thalamic tract. *See* Tract
Medulla oblongata
  blood supply, 243
  development, 9
  external features, 30–34
  internal structure, 35–39
Medullary
  lamina, internal of thalamus, 92
  velum
    inferior (posterior), 138
    superior (anterior), 145
Membrane, choroid of cerebral hemisphere, 158
Mesencephalic nucleus of trigeminal nerve. *See* Nucleus
Mesencephalon. *See* Midbrain
Metathalamus, 62, 90
Midbrain
  blood supply, 226, 236
  development, 8
  external features, 49–53
  internal structure, 52–58
Monticulus
  culmen of, 133
  declive of, 133
Motor area of cortex. *See* Area, of cerebral cortex
Motor
  areas, extrapyramidal, 180, 181
  ganglion (autonomic), 4, 208
  nerves, 4
  nucleus, 103

NEOCEREBELLUM, 149
Nerve
  abducens, 101, 106
  accessory, 103, 110, 111
  acoustic, 102
  auditory, 47
  cardiac, 216
  chorda tympani, 218
  ciliary, 218
  cochlear, 47
  cranial, 100–124
  definition, 7
  eighth cranial, 102
  eleventh cranial, 103
  facial, 101
    motor
      branchial, 112
      preganglionic, 109
    sensory
      general, 115
      root, 112
      taste, 116
  fifth cranial, 101
  first cranial, 100
  fourth cranial, 101
  glossopharyngeal, 102
    motor, 109, 110
    sensory, 116
  hypoglossal, 103, 106
  intermedius, 112
  ninth cranial, 112
  oculomotor, 100
    somite, 108
    preganglionic, 108, 110
  olfactory, 100
  optic, 66, 100
  parasympathetic, 209, 217
    afferents, 220
  petrosal
    superficial, 218, 219
    deep, 218
  postganglionic, 5
  preganglionic, 5
    sympathetic, 18, 209
    parasympathetic, 217–220
    of pterygoid canal, 218
  second cranial, 100
  sensory, definition, 4
  sixth cranial, 101
  seventh cranial, 101
  spinal, roots of, 14
  splanchnic, 213
  sympathetic, 209–215
    afferents, 217
  tenth cranial, 102
  third cranial, 100
  trigeminal, 101
    motor, 112
    sensory, 34, 115
  trochlear, 101, 108
  twelfth cranial, 103, 106
  vagus, 102
    motor
      preganglionic, 108
      branchial, 110
    sensory
      general sensory, 115
      visceral, 116
  vestibular, 102, 117
Nerve fibre
  definition, 2
  sensory, 4
Nerve impulse, 2
Nervous system
  anatomical unit, 1

INDEX 253

autonomic, 207–220
cavity of, 10, 11
central part, subdivisions, 8, 9
functional unit, 2
parasympathetic, 209, 217–220
  hypothalamic centres, 86
peripheral part, composition, 4–6
sympathetic, 209–216
  hypothalamic centres, 86
Neocerebellum, 149
Neocortex, 167
Neurohypophysis, 89
Neuron, 1
Neuron chain, 2–4
Nodule of cerebellum, 137, 143
Nucleus
  of abducens nerve, 106
    blood supply, 238
  accessory cuneate, 27
  of accessory nerve, 110
  ambiguus, 110
  amygdaloid, 202
    connections, 203
    relations to uncus, 179
  anterolateral of cord, 17
  anteromedial of cord, 17
  of Burdach. See Nucleus, cuneatus
  caudate, 201, 202
    blood supply, 232, 233
  central of cord, 17
  of cerebellum. See Cerebellum
  cochlear, 47, 121
  of cranial nerves, 103–121
    branchial motor, 110–112
    classification, 103
    location in sections, 104
    preganglionic, 108–110
    relationships in medulla, 121
    sensory, 112–114
    somite motor, 106–108
  cuneatus, 36, 114
  definition, 7
  dentate, 141, 148
    blood supply, 226
  dorsalis, 18, 26
  Edinger-Westphal, 110
  emboliform, 141, 148
  of facial nerve
    branchial motor, 112
    preganglionic, 108, 109
    sensory, general, 115
    taste, 116
  fastigial, 142, 147, 148
  globose, 142, 147
  of glossopharyngeal nerve
    branchial motor, 110
    preganglionic, 109
    general sensory, 115
    visceral sensory, 116
  gracilis, 36, 114
  of hypoglossal nerve, 106
  of hypothalamus, 82, 84, 85
  of inferior colliculus, 51, 56
  intermediate of cord, 17
  intermediolateral of cord, 18
  lentiform, 68, 199, 201, 202
  of Luys. See Nucleus, subthalamic
  of mammillary body, 85
  of oculomotor nerve, 108, 110
  olivary
    inferior (olive), 33, 38, 144
    superior, 47, 48, 121
  paraventricular, 85, 89
  pontine, 42, 48
  red, 57, 79
  salivary
    inferior, 109
    superior, 109
  sensory, in medulla, 121
  of spinal cord, 16–18
  of spinal tract of trigeminal nerve, 37, 112, 113
  subthalamic, 79
  supraoptic, 85, 89
  of thalamus
    anterior, 99
    anterior ventral, 95
    centromedian, 98
    dorsolateral, 97
    dorsomedial, 98
    lateral, 94
      efferent fibres, 97
    lateral ventral, 95
    medial, 97–99
    posterior, 95
    posterior ventral, 95
    posterolateral, 97
  of tractus solitarius, 112, 113
    afferents, 116
    efferents, 117
  of trigeminal nerve
    main sensory, 113
      connections, 115
    mesencephalic, 113
      connections, 115
    motor, 112
    spinal, 37, 112, 113
      connections, 115
  of trochlear nerve, 108
  of vagus nerve

branchial motor, 110
preganglionic, 108
general sensory, 115
visceral sensory, 115
vestibular, 112–114
connections, 117–121

OCULOMOTOR NERVE. See Nerve
Olfactory apparatus
nerve, 100
bulb, 154, 166
tract, 156, 166
trigone, 166
lateral olfactory stria, 166
lateral olfactory gyrus, 166
sensory area of cortex, 180
pathways, 165, 166, 188
Olive, 33
Opercula, 157
Optic apparatus
nerve, 100
chiasma, 66, 74
tract, 66
radiation, 90
blood supply, 233, 237
cortical projection, 180
reflex connections, 52, 120

PAIN
in disease of thalamus, 242
pathway
in cord, 22
in brain, 115
referred from viscera, 216, 217
Paleocerebellum, 149
Pallidum, 203
Papez, emotion, 189
Paraflocculus, 136
Paralysis, spastic, 125
Paramedian lobule. See Lobule, gracilis
Parasympathetic nervous system, 209, 217–220
hypothalamic centres, 86
Parolfactory area, 179
Pathway
auditory. See Auditory apparatus
definition, 6
cortico-ponto-cerebellar
development, 42
cortical origin, 72
in internal capsule, 72
in diencephalon, 72

in midbrain, 51
in pons, 48
in cerebellum, 139
dento-rubro-thalamic
dentate nucleus, 141
brachium conjunctivum, 141, 44, 58
in subthalamus, 79
in lateral ventral nucleus, 95
extrapyramidal
via reticular formation
premotor area, 181
in internal capsule, 182
in midbrain, 182
in spinal cord, 124
via brachium of superior colliculus
occipital extrapyramidal area, 182
in internal capsule, 92
in diencephalon, 92
in midbrain, 52, 56
below midbrain, 57
general sensory, 79, 97
intersegmental, 21
olfactory. See Olfactory apparatus
optic. See Optic apparatus
pain and temperature
in cord, 22, 23
in medulla, 39
in pons, 47
in midbrain, 58
in subthalamus, 79
in thalamus, 95
in internal capsule, 97
cortical projection, 179
position sense
in cord, 24
in medulla, 36
in pons segment, 47
in midbrain, 50
in subthalamus, 79
in thalamus, 95
in internal capsule, 97
cortical projection, 179
pyramidal (corticospinal, corticobulbar tracts)
motor cortex, 180, 181
in internal capsule, 72
in diencephalon, 72
in midbrain, 51
in pons segment, 48
in medulla, 31, 35
in spinal cord, 28, 29
interruption in internal capsule, 182
segmental, 21

spinocerebellar
  dorsal, 26, 33
  from arm, 26, 33
  ventral, 25, 33
  temperature. See Pathway, pain and temperature
  touch
    in cord
      crossed, 24
      uncrossed, 23
    in medulla, 37
    in pons segment, 47
    in midbrain, 50
    in subthalamus, 79
    in thalamus, 95
    in internal capsule, 97
    cortical projection, 179
  vestibular. See Vestibular apparatus
  visual. See Optic apparatus
Peduncle
  cerebellar
    inferior. See Restiform body
    middle. See Brachium pontis
    superior. See Brachium conjunctivum
  cerebral, 53
  inferior thalamis, 98, 99, 202
Perforated areas. See Area, perforated
Pineal gland, 66
Plexus sympathetic, 213
Poles, of cerebral hemisphere, 163
Pons segment
  blood supply, 242
  development, 41–43
  external features, 44–46
  internal structure, 46–48
  subdivisions
    basilar, 44, 48
    tegmentum, 44, 46–48
Position sense. See Pathway
Postganglionic nerve, 5
Precuneus, 179
Prefrontal cerebral cortex, 184
Preganglionic nerve, 5
Premotor area, 181
Preoptic region, 85
  recess, 74
Pretectal region, 64, 95
Projection fibres. See Fibres
Proprioceptive pathways. See Pathway, position sense; Tract, spinocerebellar
Pulvinar, 66, 92
Pupillary constriction, 64, 75
  dilation, 87
Purkinje cells, 146, 147, 148
Putamen, 201, 202
Pyramidal pathway. See Pathway
Pyramis, of cerebellum, 136

RADIATION
  auditory, 92
  optic, 90, 233, 237
Ramus (Rami) communicantes
  grey, 212
  white, 212
Recess
  dorsal of fourth ventricle, 45
  lateral of fourth ventricle, 38
  preoptic, 74
  suprapineal, 75
Referred pain, 217
Reflex pathways
  intersegmental, 21
  segmental, 21
Relay station, 7
Restiform body
  in medulla, 33
  in stalk of cerebellum, 44
  in cerebellum, 139
Reticular formation, 123–125
  cerebellar connections, 124
  reticulospinal tracts, 124
  reticulothalamic tract, 124
  role in spastic paralysis, 125
Retina
  layers, 100
  cortical projection, 180
Rhinal sulcus, 178
Rigidity, 58, 81
Root
  of accessory nerve, 110, 111
  dorsal and ventral, 14, 19
  sensory of facial, 112
Rostrum of corpus callosum, 161

SALIVARY GLANDS, innervation of, 218, 219
Salivary nuclei, 109
Segmentation of spinal cord, 14
Sensory
  area of cortex, 169
  fibres, 4
  ganglion, 4
  nuclei, 103
Septal area, 158, 188
  clinical concept, 179
Septum pellucidum, 163, 206

Sleep, 124
Sommer's sector, 235
Spasticity, 125, 182
Speech, cortical areas, 184
   dominant hemisphere, 184
Spinal cord, 13–29
   blood supply, 238, 244
   canal, 10
   external features, 13
   internal structure, 15–19
   pathways, 20–29
Spinocerebellar tracts. See Tract
Splanchnic nerves, 213
Splenium of corpus callosum, 161
Stem of posterior cerebral artery, 223
Stalk of hypophysis, 89
   of pineal gland, 75
Stria
   lateral olfactory, 166
   longitudinal, 188
   medullaris thalami, 99
   terminalis, 203
Striatum, 203
Subarachnoid space, 11
Subcallosal gyrus (area), 178
Substance (Substantia)
   gelatinosa, 16
   innominate, 201, 203
   nigra, 57, 79
   perforated
      anterior, 157, 199
      posterior, 236, 242
   reticular, 123–125
Subthalamus, 62, 77–82
   relationships, 77
   internal structure, 79
   nuclei, 79
   fibre bundles, 81–82
Sulcus
   calcarine, 177
   callosal, 177
   central, 171
   cerebral
      definition, 167
      evolution, 174
      relation to areas, 170
   cinguli, 177
   collateral, 178
   frontal, 172
   hippocampal, 177
   hypothalamic, 76
   intraparietal, 174
   lateral, 171
      development, 157
   limitans, 38, 46

   parieto-occipital, 172, 177
   postcentral, 174
   precentral, 172
   rhinal, 178
   temporal, 174, 178
Supraopticohypophyseal tract, 88.
      See also figure 54
Suprapineal recess, 75
Sympathetic nervous system. See
      Nervous system
Synapse, 2
Syndrome
   Horner's, 87
   thalamic, 242

TAPETUM, 196
Tectobulbar tract, 57
Tectocerebellar tract, 145
Tectospinal tract, 57
Tectum, 53
Tegmental decussations, 56, 57
Tegmentum
   midbrain, 54
   pons segment, 44, 46–48
Telencephalon, 8
Temperature
   pathway. See Pathway
   regulation, 86
Thalamic
   fasciculus, 81, 95
   nuclei, 94–99
   syndrome, 242
Thalamocortical connections, 184
Thalamus, 92–99
   blood supply, 240
   form, relationships, 92
   function, 62
   internal structure, 92
   nuclei, 94–99
Tonsil of cerebellum, 136
Touch pathway. See Pathway
Tract
   corticobulbar. See Pathway,
      pyramidal
   corticopontine. See Pathway, cortico-
      ponto-cerebellar
   corticospinal. See Pathway, pyramidal
   definition, 7
   fastigiobulbar, 140, 148
   frontopontine. See Pathway, cortico-
      ponto-cerebellar
   geniculocalcarine. See Radiation,
      optic
   of Goll. See Fasciculus gracilis
   of Lissauer, 19

mammillo-tegmental, 85, 87
mammillo-thalamic, 85, 87, 99
olfactory, 156, 166
olivocerebellar, 144
optic, 66
pontocerebellar, 144
pyramidal. See Pathway, pyramidal
resticulospinal, 124
reticulothalamic
   in midbrain, 124
   in subthalamus, 79
   in thalamus, 98
rubroreticular, 55, 57
solitarius, 116
spinal tract of trigeminal nerve, 34, 115
spinocerebellar
   dorsal, 26, 33, 144
   ventral, 25, 33, 144
spinothalamic, lateral
   in cord, 23
   in medulla, 39, 115
   in pons segment, 47
   in midbrain, 58
   in diencephalon, 79, 95
spinothalamic, ventral
   in cord, 24
   in medulla, 37
supraopticohypophyseal, 88. See also figure 54
tectobulbar, 57
tectocerebellar, 145
tectospinal, 57
temporopontine. See Pathway, cortico-ponto-cerebellar
vestibulocerebellar, 118, 144
vestibulospinal, 120
Transverse cerebral fissure, 159
Trigeminal nerve. See Nerve
Trigone
   habenular, 66
   olfactory, 166
   vagal, 108
Trochlear nerve. See Nerve

Trunk, sympathetic, 212
Tuber cinereum, 68, 74
   of vermis, 135, 145
Tubercle (Tuberculum)
   anterior of thalamus, 66, 92
   cinereum (medulla) 34
   cuneate, 31

UNCINATE FASCICULUS, 193
Uncus, 177, 179
Uvula, 136

VAGUS NERVE. See Nerve
Vallecula, 132
Velum
   inferior (posterior), 11, 32, 42, 138
   superior (anterior), 11, 42, 132
Ventricle
   definition, 10
   fourth, 10, 37
   lateral, 11, 156, 203–206
   third, 11, 72–76
Vermis, 128
Vestibular apparatus
   nerve, 102, 117
   ganglion, 102
   nucleus, 112–114
   secondary pathways, 117–121, 144
Viscera, innervation of, 213, 217
Visceral afferent fibres
   in sympathetic nerves, 216
   in parasympathetic nerves, 220
Visual apparatus. See Optic apparatus

WAKEFULNESS AND RETICULAR
   FORMATION, 124
Water balance and hypothalamus, 89
White commissure, 19, 23, 24
   matter, 7
   rami communicantes, 212
Word blindness, 184
   deafness, 184

ZONA INCERTA, 80

www.ingramcontent.com/pod-product-compliance
Lightning Source LLC
Chambersburg PA
CBHW020249030426
42336CB00010B/691